Fossil Animal Remains: their preparation and conservation

by
A. E. RIXON

THE ATHLONE PRESS *of the University of London* 1976

Published by
THE ATHLONE PRESS
UNIVERSITY OF LONDON
at 4 *Gower Street, London* WC1

Distributed by Tiptree Book Services Ltd
Tiptree, Essex

U.S.A. and Canada
Humanities Press Inc.
New Jersey

© *University of London* 1976

ISBN 0 485 12028 3

Text set in 12 pt Photon Imprint,
printed by photolithography, and bound
in Great Britain at The Pitman Press, Bath

Preface

During the greater part of my career as a preparator it was part of my duties to answer enquiries concerning the preparation and conservation of fossils. They came from a wide spectrum of society; ranging from the schoolboy amateur to the museum curator and university research worker. It was suggested to me by some of these people that it would be useful if at sometime I should write in book form the answers to most of these queries. This I have tried to do in a simple and understandable way.

Because it is intended as a practical manual, this book, like Gilbert's House of Peers, makes 'no pretence to intellectual eminence or scholarship divine' and theory has only been dealt with in those places where it is necessary to the understanding of the technique described. Although, in my time, I have prepared almost every type of fossil there is, it will be apparent from the contents of this work that my main interest has been in the vertebrates.

I have said but little about fossil plants; because in the last forty years the science of palaeobotany has developed its own vast technology which would require a separate book to describe fully, written by a person with far more knowledge in this field than I possess.

It will be noted that reference is made in several places to the work of archaeologists and art restorers. Through membership of the International Institute for the Conservation of Historic and Art Objects I have had a long, happy and rewarding connection with people in these fields and I owe much to their influence.

Many of the techniques described are not new and where I have been able to find the originator of a method I have ascribed it to him. If through ignorance I have in any instance failed to do this, I offer my apologies to the offended party and assure him that no offence was intended. Since I have now retired from the profession any opinions expressed in this book are my own and do not necessarily reflect those of any other person or institution.

My especial thanks are due to my wife, Pat, who typed the manuscript several times, drew the diagrams and read the book to make sure that what I had written would be understood by anyone with no prior knowledge of the subject. Without her aid and encouragement it would probably never have been written.

I am grateful to Mr and Mrs A. M. Wood and to Dr J. Attridge for many helpful suggestions for the improvement of the book.

Acknowledgements and thanks are due to the Trustees of the British Museum (Natural History), Messrs Imitor Ltd, Messrs Burgess Power Tools Ltd, and to Messrs Desoutter Bros. for permission to publish photographs which are their copyright.

A.E.R.

Contents

1 The Function of the Palaeontological Laboratory and its Staff	1
2 The Consolidation and Repair of Specimens	5
3 Collecting and Treatment in the Field	36
4 Mechanical Development	59
5 Chemical Methods of Development	84
6 The Concentration of Fossils from Bulk Matrices	123
7 The Effects of the Decomposition of Iron Pyrites within a Specimen and Methods used for its Arrest	139
8 Mounting Fossils for Exhibition	153
9 Casting	193
10 The Equipment and Management of an Ideal Laboratory	230
11 Some Chemicals and Natural Substances used in Palaeontological Techniques	237
Appendix 1: Special Tools and Techniques	253
Appendix 2: Materials, Manufacturers, Suppliers	275
Index	283

Warning

Due to recent publicity concerning the potential long-term health hazards involved in the use of asbestos it is recommended that wherever it is suggested in this book it should be replaced by the substitutes listed below.

Chapter 9
(p. 202) *Contents of lower half of double saucepan used to melt P.V.C. compounds.* Use sand or the type of glass wool made for insulating lofts instead of asbestos wool.
(p. 204) *Heat resistant putty for use with P.V.C. thermosetting pastes.* Use heavy kaolin in place of asbestos powder. This was the original formula but because it has a limited useful life the change was made to asbestos.

Appendix I
(p. 254) *Forging and Tempering Cold Chisels.* Replace *asbestos blocks* in forge with broken fire bricks. Do not use *asbestos glove*. A cotton or linen rag soaked in cold water will do.
(pp. 256–7) *Two Pieces of Apparatus for use with Polyethylene Glycol 4000.* Construct box for melting pot from hardboard which will not ignite easily even if held in a bunsen burner flame for several minutes. Cover handle of saucepan with the same material.
(p. 262) *A Small Heating Tool.* Use woven glass tape (obtainable from suppliers of glass fibre materials) in place of asbestos paper to cover element.

1
The Function of the Palaeontological Laboratory and Its Staff

The role of the staff of a palaeontological laboratory is the preparation and conservation of fossils for the purposes of research by scientists, exhibition in public galleries or storage in a study collection. The word 'preparation' has been used traditionally to describe a variety of operations ranging from the consolidation and repair of fossils to their extraction from the matrix rock and their final mounting for museum display. Accordingly, a person engaged in this work is universally known as a 'preparator', although his official title may vary from institution to institution and he may be called a Research Assistant, a Conservation Officer or an Experimental Officer.

Early workers in this field were limited in their scope by the materials and tools available to them. The invention of man-made plastics and the introduction of small power tools, together with the discovery that certain fossils can be developed by chemical methods has made it possible for the modern preparator to produce results which could not be achieved previously. At the same time he must now understand the properties of a large range of synthetic resins and organic solvents used in the making of consolidating solutions and adhesives and of chemicals which will dissolve or disintegrate matrices without damaging the fossils contained within them.

The solvents used to make modern consolidating solutions and adhesives are mostly inflammable and some are toxic if inhaled in sufficient quantities; which makes it vitally important that all members of the staff should be made aware of these hazards and instructed in methods by which they may be minimised. Sheer common sense dictates that the laboratory must be well ventilated, must be provided with fire extinguishers capable of dealing with all types of fire, including those involving electrical equipment, and there must be a fire exit. A copy of a reference

2 The Palaeontological Laboratory

book, such as the *Merck Index*, which lists the properties of all the solvents likely to be encountered should be provided and every member of staff should be encouraged to refer to it.

Consolidants and adhesives may be needed for use in the field, for some specimens are friable and need to be strengthened before they can be removed from the beds in which they are found; even then they may be so fragile that they require very careful packing before they can be conveyed to the laboratory. Preparators are usually sent into the field to collect large specimens which may be found in either hard or soft rocks. Depending on the type of rock and the specimen, these fossils may need either to be extracted from the beds encased in blocks of matrix, or most of the matrix may be removed on the spot. In either case it may be necessary to encase the specimens in a protective coating of plaster bandages or plastic foam to protect them in transit.

On their arrival at the laboratory the preparator will remove the field packing, preserve the field data and clear the matrix from the specimen when this has not already been done, using either mechanical or chemical methods. When he has revealed as much detail as possible and the specimen is strong enough to be handled he will pass it to a palaeontologist who is an expert on the group of animals concerned. Should the palaeontologist decide that the fossil is of sufficient general interest he may ask the preparator to mount it for exhibition or to make casts of it for distribution to scientists working in other institutions. Occasionally the laboratory staff may be called upon to design and make a support or a container for a specimen which will be at special risk in storage.

Mechanical development is carried out with hammers, chisels, small power percussion tools, various types of grinding tools, ultrasonic apparatus and dental tools. The majority of these techniques result in the production of a great deal of finely divided dust which, unless it is guarded against, is a health hazard. Respirators, conforming to the health regulations in industry, should be issued to every person doing this work and they should be urged to clean and sterilize them at regular intervals, also to change the filters before they become blocked with fine particles.

Chemical development has already been mentioned. A few of the chemicals used are not corrosive or volatile, but most preparations of this type are carried out with acids which emit corrosive fumes that are at least unpleasant and can be harmful to staff and expensive equipment such as microscopes; ideally they should not be used outside a fume cupboard. Where one is not available this type of preparation should be carried out under a lean-to shed in the open air. It should never be attempted in a normal dwelling house and is therefore unsuitable for use by most amateur palaeontologists. Acid development has revolutionised the research into many types of vertebrate fossils for, where used correctly, it can reveal the smallest details in the anatomy of a specimen, thus making available to the scientist all the information he requires. Before its discovery many structures remained covered in matrix because it was impossible to clear them by mechanical means without destroying the specimen.

The rocks from some localities are packed with small fossils and when there is a chemical difference between the matrix and the specimen, lumps of the material may be broken down with chemicals. After the solution is complete the fossils may be separated from the insoluble fractions of the matrix by various means; the preparator must decide which is the most efficient method to use in every case.

Fossils collected from beds containing much finely divided iron pyrites are often damaged by the decay of this chemical and the preparator must know how this, in part, may be avoided or arrested. No known method is absolutely reliable and there is room for research into this problem. Other causes of damage to fossils in storage are the presence of salt or natural soda in the cells of the specimen. Should specimens containing these chemicals be stored at too high a humidity or become wet in any other way, these salts will dissolve and, on drying, recrystallise causing damage.

All the techniques mentioned above must be understood by a worker in a palaeontology laboratory and they are described in full in the chapters of this book. It should have emerged from what has been said above that a preparator needs a working knowledge of many things. He must be part chemist, part anatomist and part artist, added to which he must be capable of working in a variety of materials ranging from all forms of plastic to mild steel. He is a living contradiction of the old adage, for he must be a jack of all trades in order to be the master of his own; but the most essential piece of knowledge he must have is an awareness of his own limitations. No one can be an expert in everything so when confronted with a problem which is outside his experience, he must never guess but consult an expert or read up the subject in text books.

The broad spectrum of the work of a preparator has now been described and it only remains to say something about the laboratory in which he performs his duties. It is not possible to be specific about the structure and equipment of a laboratory as this varies according to the needs of the institution which it serves. Most universities have scientists doing research work on fossils and they may be adequately catered for by one man working in a small room. At the other end of the scale the large national museums in many countries may employ anything from six to a dozen preparators and they obviously require more room and bench space. It is only when a new building is being planned that an ideal laboratory can be designed, so most preparators have to make the best of what is available. Chapter 10 of this book describes the type of laboratory that every preparator would like to have but is unlikely to get, unless it is in a new building; however it contains some suggestions which could be used to improve an existing laboratory without the need of major structural alteration.

2
The Consolidation and Repair of Specimens

The physical condition of fossils varies according to the beds in which they are found, their chemical composition and their cellular structure. Some are robust and may be collected and extracted from the matrix rock without any treatment, while others are friable or comminuted and need to be strengthened before they can be safely handled. The process of strengthening specimens in a weak condition is known as 'hardening' or consolidation and consists of the introduction of solutions of various adhesives, in appropriate solvents, into the pores and cracks in the fossil so that when the solvent has evaporated, the crumbling or loosely attached fragments are stuck together by the solid component of the solution. To be effective, the solution must penetrate deeply and deposit as much of the adhesive as possible inside the specimen. Because the porosity and properties of absorbtion vary from fossil to fossil and the optimum strength of a consolidating solution is that having the highest viscosity which will achieve the necessary penetration, it is only in exceptional cases that the proportions of solid to liquid can be given. The only way to overcome this problem is to make stock solutions of the consolidants in general use, of a viscosity far in excess of normal needs and to dilute them to suit the condition of each individual specimen. The number of applications of a consolidant required to make a fossil strong enough to survive also varies from specimen to specimen, but in all cases where more than one is needed, each must be allowed to dry before the next is applied. Should a fossil requiring consolidation also be broken, each separate piece must be consolidated before any repair is attempted and in cases where its original condition was very friable, an adhesive based on a solvent which will not dissolve the consolidant should be used.

As stated above, consolidating agents are weak solutions of adhesives and many of the same solids, dissolved in small quantities of solvent to make very viscous solutions are used to repair fossils. In a few cases the materials, when employed as an adhesive, are applied in a molten state, but this practice is now less common than it was before man-made resins were available. In order to avoid repetition, when any of the substances described below have the dual function of consolidant and adhesive this will be mentioned in the description of their properties, although the techniques of consolidation and repair will be dealt with separately.

The practice of consolidating and repairing fossils is now more than a century old and many different substances which are no longer considered efficient have been used in the past, and so it is important, for several reasons, that preparators working in institutions which have old collections should know what they are and how to recognise them. Some decay with time and others are water soluble or will swell in water, making special precautions necessary if for any reason specimens impregnated with them have to receive wet treatment. Others are not compatible with modern consolidants and adhesives and it is necessary to remove them before broken specimens containing them can be repaired. Those which were in most common use are discussed below, and again where they had the dual function of consolidant and adhesive this will be mentioned in the description of their properties.

Substances used for Consolidation and Repair prior to 1940

Animal Glues and Gelatines

Used at normal strength as adhesives, and thinned with water as consolidants, they were applied hot from a carpenter's glue pot. The simplest test for them is to remove a piece of the adhesive and burn it, either in a pair of tongs or a hard glass tube, when it will blacken and give off a very characteristic smell resembling that of burnt bone. It is normally possible to obtain a big enough sample for this purpose as the material was usually applied in excess. When this cannot be done a small fragment of the specimen should be soaked in hot de-ionised water for one hour. The liquid is then filtered off and made strongly alkaline with a solution of sodium hydroxide. To this is added, drop by drop, a solution of copper sulphate, and if the solution changes colour to pink or violet the presence of glue or gelatine is established. This is the Biuret test.

Fortunately, glue is not soluble in most organic solvents and is compatible with emulsions of polyvinyl acetate, so these can be used where it is present.

Glue can do the greatest damage when much too much has been applied and on drying, over a period of many years, it has contracted and curled up, breaking itself into fragments and pulling the surface of the bone with it. Owing to their structure this is most likely to happen to avian bones. The treatment which is long and difficult and not always successful, is discussed in the section dealing with repair.

Where breaks alone are involved glue-treated specimens can be repaired with thick polyvinyl acetate emulsion or almost any of the adhesives to be mentioned later.

It is likely that any specimens which have been in a collection since before the 1920s have either been consolidated or repaired with animal glue in one form or another and this should be taken into account if acid development or any other wet process is contemplated. Should the presence of glue be established it must be decided whether it can be washed out without damaging the specimen and the fossil treated with a suitable substitute.

In all such cases the person responsible for the specimen should be informed before any action is taken, and even if in consultation with him, it is decided that a calculated risk is justified nothing should be done until a good cast has been made and photographs have been taken. Any action involving the soaking out of glue joints or the removal of glue from a specimen must be kept under close observation and carried out with great caution. Given care, it is unlikely that a situation will arise which cannot be dealt with by a competent preparator.

Shellac

This is a resinous substance produced by an insect (*Coccus lacca*) and a solution in methylated spirit was at one time in almost universal use as a consolidant. When used in this way it gives little trouble, but there appears to have been a time when it was widely used as an adhesive too. From the appearance of most joints made with this substance it was applied in a molten state, and partly allowed to solidify before the joint was made. This was presumably done to form a gap-filling paste. Such joints are very brittle and the re-making of them is a common task.

Shellac is easily recognised by its reddish brown colour and the very characteristic odour of burnt French polish it gives off if touched with a hot iron. It is fairly easily removed from a joint by mechanical means and any residue may be dissolved in industrial methylated spirit or other alcohols. When the resin is very old it may be necessary to add a small amount of ammonia solution to the alcohol. Joints may be re-made with almost any modern adhesive.

Waxes

These are the most difficult of old materials to deal with. Almost every type of natural wax seems to have been used, either alone or in mixtures with others. Paraffin wax does not appear to have been in favour. Fortunately, except for elephant's teeth and tusks, waxes were rarely employed as consolidants, but were widely used as adhesives and sometimes for restoration purposes. The most common wax used as a cement is yellow and brittle; this is probably due to a high content of carnauba wax. Most of it can be scraped off a joint or picked off with needles. When the fossil is porous the residue is best removed by applying a pad of surgical cellulose soaked in methylene chloride to the joint surface and covering this with polythene sheet firmly held to it and fixed with rubber bands to make an impervious outer coat. The whole is left to stand for several hours. The polythene cover is then removed, and making sure that the methylene chloride pad is still tight to the surface, the specimen is placed in a fume cupboard and the solvent allowed to evaporate completely.

When this has happened the wax will have migrated from the fossil to the outer surface of the pad. The process may have to be repeated. If all the wax has been removed the joint can be re-made with any adhesive; should only a small amount remain, polyvinyl acetate emulsion will usually make a good joint. The above process may be used on all wax joints.

It is of course useless to attempt to repair an old joint, from which the wax has not been removed, with an adhesive which contains a wax solvent, because wax will dissolve in the solvent and will form what is effectively a paint stripper and this will not dry for a very long time.

Cellulose Nitrate

This substance in the form of celluloid was widely used up until the 1940s, and is still found in some proprietary adhesives. A thin solution made in a mixture of acetone and amyl acetate was used for consolidation and a more viscous solution in the same solvents as an adhesive for small joints. Joints made with it embrittle in time and sometimes sheets of it will tear loose from the surface of non-porous fossils to which it has been applied. It burns with almost explosive violence giving off a characteristic odour. It is soluble in both ketones, such as acetone and methyl ethyl ketone, and in esters like amyl acetate and n-butyl acetate. Cellulose nitrate is easy to remove from failed joints as it normally pulls off when the joint breaks and can be torn away.

Cellulose Acetate

Owing to the fact that this is soluble in ketones like acetone but is not soluble in alcohols and the esters, it still has its uses as a temporary supporting medium. Unlike cellulose nitrate it does not burn, but shrivels and blackens in a flame.

Modern Consolidating Agents and Adhesives

These are usually man-made plastics dissolved in various solvents or emulsions in water. As to efficiency, there is little to choose between most of them and choice is very often dictated by personal preference and availability. Those discussed below are in most common use in the British Isles. Workers in other countries have published papers in which proprietary reagents, which are not easily obtained here, are mentioned and since the substances which compose these consolidants and adhesives are rarely mentioned it is not possible to give a fair evaluation of them.

Consolidation and Repair of Specimens

Polyvinyl butyrals

As the basis of both consolidants and adhesives these are among the best resins obtainable. Those most widely used in the British Isles are the 'Butvars' manufactured by Monsanto Ltd. Butvar B98 dissolved in isopropyl alcohol is an excellent consolidant, and B76, the only member of this range that is soluble in acetone, makes a good fast drying adhesive. Cairn Chemicals Ltd sell polyvinyl butyral M150 (soluble in acetone and diacetone alcohol) in small quantities to museums. In Germany, Farbwerke Hoechst A.G. manufacture polyvinyl butyrals under the trade name 'Mowital'. Solutions of all these resins are colourless.

Polyvinyl Acetate

This has been in use for more than forty years and is a reliable substance. For consolidation purposes it is made into a solution in either toluene or industrial methylated spirit. The only objections to it are that, when it is allowed to dry in air, it imparts a high gloss to specimens which is unacceptable in some institutions, and that some grades rapidly become sticky if the object is handled for long. The gloss is reduced if the specimen is dried slowly in an atmosphere of the solvent in which the plastic was dissolved. Small specimens may be placed under a bell jar together with a small open vessel of the solvent and left to dry, larger ones can be covered with polythene sheet supported on a frame.

A thick solution in ethyl acetate or acetone makes a good adhesive which dries very quickly. Polyvinyl acetate is also made in the form of emulsions and certain grades are of great and varied use in the repair and preparation of fossils. The emulsion chosen should have a high plasticiser content and the particles be as small as possible. A grade such as Vinamul N 9146, manufactured by Vinyl Products Ltd, is excellent for these purposes. It can be used to consolidate and strengthen waterlogged specimens, and is of particular use in the treatment of elephant tusks and teeth and deer antlers. The emulsion, as provided by the makers, is too viscous for this purpose but is an excellent adhesive which is very clean and easy to use. This material is the basis of emulsion paints and of some commercial adhesives, so a suitable type should be obtainable in most countries.

Polybutylmethacrylate

In the form of a solution in toluene called Bedacryl 122x, supplied by I.C.I. Ltd, this has been used for over twenty years in the consolidation of, and as a protective coat for, fossils which have been damaged by the breakdown of pyrites. It has sometimes been used as a general consolidant. The objections to it are as stated for polyvinyl acetate and the cure is the same.

In acid development of fossils it is the most effective and most widely used substance for strengthening the specimen. For this purpose products called Vinalak 5909 or 5911, made by Vinyl Products Ltd, are normally used, diluted with methyl ethyl ketone. These plastics will withstand long immersion in dilute acetic acid and formic acid, as well as in water. Methyl ethyl ketone is the diluent preferred by most workers, but acetone could be used or in the absence of both, toluene will do, but takes longer to dry. The solution again imparts a shiny finish to the specimen and becomes sticky if handled too much. A recent technique, to be mentioned later, has been evolved to overcome this trouble in acid-treated specimens. The shine produced by Bedacryl and by Vinalak when used for pyritised specimens can be diminished if a small amount of aerogel silica, Santocel 54 made by Monsanto Chemicals Ltd, is suspended in the solution. When this is done care must be taken to see that the solution is not allowed to collect in pools in hollows in the specimen or it will dry out white in these areas. Aerogel silica is used as a thixotropic agent in some forms of fibreglass resin work and it is possible that small quantities can be bought from the suppliers of epoxy resins.

Polymethylmethacrylate

Almost universally obtainable this plastic is normally used as a quick-drying adhesive in solution in either ethyl acetate or chloroform. Because ethyl acetate, although inflammable, is physiologically very much safer, this should be used. It is particularly useful as an adhesive for the repair of specimens which are to be developed in acids. Although it is soluble in glacial acetic acid, it is not in a 10 per cent v/v solution of the acid in water and in this will stand prolonged immersion. It has been used by some workers in place of polybutylmethacrylate for coating and consolidating specimens to be prepared in acids but this is not recommended because if an excess is applied it tends to shrink away from the bone and cause damage. There are commercially produced packs of the monomer and polymer treated so that they will set when the two components are mixed together, and these can be used to fill in gaps in bones to be treated with acid. Where a part of the specimen is represented only by a mould in the rock, polymethylmethacrylate is capable of producing a reasonable cast which will join to the surrounding bone owing to the plastic's own adhesive properties. North Hill Plastics produce this type of pack in three colours, clear, ivory and black.

Polymethylmethacrylate is sold under different trade names in all countries, and can often be bought as scrap. In the British Isles it is known as 'Perspex' and is made by I.C.I. Ltd. In America it is called 'Lucite'. Much time may be saved when making solutions of this material if it can be obtained in the powder form used for injection mouldings.

Polyvinyl Alcohol

The main attraction of this plastic is that it is soluble in practically nothing but water and is very resistant to the organic solvents most commonly used in palaeontology laboratories. A viscous solution may, therefore, be used as an adhesive for small joints which are likely to become flooded with organic solvents in some other part of the preparative operation; or when normal consolidating agents have to be removed from a specimen, a more mobile solution can be employed to strengthen the object after one side has been cleaned. It is applied and allowed to dry before the specimen is turned over to remove the plastic from the other side. A solution of polyvinyl alcohol can with advantage, replace mucilages of gum tragacanth for sticking small specimens into cavity cells, etc. The plastic is sold by most chemical suppliers as a very fine powder and it is necessary to make it into a lump-free paste by mixing it with a little water, much as table mustard is made, before attempting to make it into a solution.

Soluble Nylon

First introduced for use in archaeological conservation by Werner (1958) this plastic has been found to have applications in some palaeontological techniques. Its great advantage is that it is hardly discernable after it has been applied. While it is not water soluble, it is water permeable and salts may be washed from a specimen to which it has been applied. Care, however, is needed if only temporary application is required, for exposure to even moderately high temperatures or organic acids has been found to affect its solubility and make it difficult to remove. Its greatest use has been in the treatment of chalk echinoderms and other light coloured invertebrate fossils.

The plastic is used as a 5 per cent solution in a 1:10 v/v mixture of de-ionised water and industrial methylated spirit. Solution is quicker if carried out in a water bath at about 60°C. Methanol is a good solvent for this material. A 10 per cent solution is a gel at room temperature, but, if the temperature of the solution is raised slightly, it becomes a very mobile liquid. Methanol is both inflammable and toxic so must be used with care. At the suggestion of Mr Baynes-Cope of the British Museum Laboratory, the author has used a solution in methanol to strengthen some extremely weak and decaying original labels and the results were excellent. Soluble nylon is manufactured in the British Isles by I.C.I. Ltd, and is obtainable in small quantities from Museum and Archaeological Suppliers.

Polyethylene Glycol 4000

This wax-like substance has all the advantages of paraffin wax and none of its disadvantages. It is widely used in archaeology and sometimes by palaeontologists for the preservation of waterlogged wood and several articles have appeared in various journals describing its use in this field. It has a low melting point and is soluble in water. Although it is not generally used to consolidate bone, some small pieces of Pleistocene antler treated with molten polyethylene glycol several years ago are still in good condition and further investigation of its uses in this field would be worth while. It can also be used as a solution in water to consolidate wet specimens. The main uses it has been put to in palaeontological work so far are as a temporary and easily removed adhesive for the arrangement of fossils for photography, for their attachment to storage mounts and as a support for small and delicate fossils requiring mechanical preparation. It has also been used in place of wires to hold the bones making up the limb of a small plesiosaur which were mounted flat for exhibition. For all these purposes it is applied in the molten state and allowed to solidify. Details of its various uses will be discussed later.

Gap-Filling Adhesives Based on Jute Floc and Various Plastics

There are two such adhesives in common use, one a patented material called 'Fibrenyl' first mentioned by Newman (1955) and the other A.K.J. dough (Cornwall, 1956). Both are, in effect, a dough made of very short staple jute floc and a solution of plastic in a mixture of solvents. The formula published by Cornwall has to be modified, as he himself notes, as the basic plastic is no longer obtainable. There is little doubt that a solution of Butvar B98 or an equivalent polyvinyl butyral 25–30 w/v in isopropyl alcohol or B76 or an equivalent at the same proportions in acetone, mixed with a roughly 3:1 mixture of short staple jute floc and heavy kaolin to make a stiff paste would be effective. The first two substances mentioned have been used as gap-filling cements for many years. Such a mixture is capable of making a strong joint in a heavy fossil. The dried material has rather a rough surface, so for restoration purposes the area should be under-filled with respect to the contours of the fossil and the outer shell built up with either plaster of Paris or Polyfilla—a commercial gap filler used in interior decorating.

At the moment jute floc is in short supply, but the dust removed from jute waste before it is processed is obtainable. This material is made up of short staple fibres and stalks. It is possible to reduce the stalks to fragments by grinding the mixture in a high speed coffee mill; of course a much larger and more robust mill is needed if this is to be done on a large scale. Several workers are attempting to find a substitute for jute in this context. Synthetic flocs such as terylene and nylon have been suggested as well as coarse sawdust and wood fibres.

Epoxy Resins and Gap Filling Cements

The epoxy resins are excellent adhesives and consolidation materials. They do, however, suffer a disadvantage for palaeontological work in that once they have set they are almost impossible to remove and some tend to discolour the specimens. Most institutions do not like using processes which are difficult to reverse but it is sometimes the case that the only satisfactory adhesive to employ for the repair of a very heavy fossil, such as a giant ammonite, is an epoxy gap-filling adhesive. The greatest advantage of the epoxy resins, apart from their strength, is the fact that there is no shrinkage as they set. Larney (1971) has said that a commercial paint stripper 'Nitromors' will dissolve out an epoxy joint of the dimensions used in the repair of porcelain. The effect of this on fossil material has not so far been tested. In the same paper he refers to a clear epoxy resin, 'Araldite' MY 790 used with hardener HY 922. This again has not been tested on fossils but might well eliminate the discolouration problem. Hempel (1969) has described the use of an epoxy of American origin called 'Maraglas' Type A655 which is colourless and shows no signs of yellowing on test. He has used this as the binding agent in a mixture which he makes to simulate marble in the restoration of statuary, and the results are excellent. With slight variations his method could, no doubt, be applied to the restoration of some fossils.

Epoxy 'plastic woods' are available and are useful in restoring missing pieces as they are easily modelled when in the plastic state and can be worked with carpenter's tools and files when they have set. The setting reaction is exothermic and this heating effect increases with the quantity used so the maker's instructions should be read for advice on this point. The usual colour is not suitable for finished work but the material is very easy to paint when it is hard.

'Permabond'

This is an alpha-cyanoacrylate monomer which almost instantly polymerises in air. There are three choices as to the speed of setting and viscosity. A joint made with this adhesive will set in seconds. Before fully set it is soluble in acetone. 'Permabond' is a new material as far as palaeontology is concerned, and because of this tests will have to be made for some time before it can be said to be absolutely satisfactory for this purpose. There is no apparent reason why it should not be and its advantages are too obvious to require stating. At first sight this material may seem prohibitively expensive, but so little is required to make a joint that in use its cost is not so high as it appears.

'Quentglaze'

Shorer (1967) first introduced this material to the world of conservation together with several others made by the same firm (see Appendix II). It is a liquid based on pre-polymers which has the property of having its cure assisted by water. This would suggest its use for consolidating wet specimens, but it is not satisfactory for this purpose because, when set, the plastic is resistant to solvents in normal use. It has been used, with great success, to consolidate wet sands or loams surrounding a fossil so that they form a part of the support when the specimen is lifted. Impressions in wet sand or soil may be preserved with this substance. 'Quentglaze' is a product which would be well worth considering for other uses in palaeontology.

Other Proprietary Adhesives

Throughout the world there are many proprietary adhesives on sale and the bulk of them are very good. It must be remembered, however, that few if any manufacturers have the repair of fossils in mind when they decide to market their product. By pure chance some of these are better for our purpose than others, and in small laboratories the use of good adhesives easily available in economic packs may be preferable to going to the trouble of making them. If this is the case the user should find out from the manufacturers upon what the adhesive is based and, if a brand name is stated in a publication, this information should be given. Should it be that for some reason or another this information cannot be obtained the preparator should at least discover by experiment what dissolves the dried adhesive and the effects of heat on it and give this information in his paper.

The important properties of any adhesive are that there should be little shrinkage on drying and in the dry state it should not be brittle. All solutions of solids have some shrinkage as they dry but emulsions have less. The expoxy resin-based adhesives are about the only common type which have none, but since their resolution is almost impossible they are only used to repair fossils in extreme cases.

Methods of Consolidation

The solutions used for consolidation may be introduced into the specimen in several ways.

Application by Brush

This is the usual method. The solution is not brushed onto the fossil directly as this may disturb loose pieces but the brush is overloaded with the consolidant which is allowed to trickle onto the specimen.

Application by Pipette

A bulb pipette made of polythene is better than one made of glass because it is impossible to break and if the end comes in contact with the fossil it is less likely to damage the specimen. Polythene tube can be bought with bores of many diameters and a tube with an internal diameter of approximately 1 cm and a wall thickness of 2 mm does very well for this purpose. When the polythene tube is heated in a stream of hot air its appearance will change from opaque white to transparent and in this condition it can be drawn out into a pipette in the same way as red hot glass. Details of the method and apparatus required for making such a pipette are given in Appendix I.

The application of a solution by pipette is to be preferred when dealing with small specimens, as the operator has more control over the amount applied than he does when using a brush. Polythene pipettes can also be used to inject solutions deep inside specimens by pushing the tube into cracks and holes.

Application by Spraying

A standard paint spraying gun may be used, but most of these have the reservoir on top so the angles at which they can be operated are limited. Cheap laboratory spray guns, powered by an aerosol can of an inert gas, are obtainable from most suppliers of chemical apparatus and have many advantages. The reservoir is below the gun and the whole apparatus is very light. The aerosol can being the only power supply needed, they may be used in the field or in other places in which operation of an air compressor is either difficult or impossible. When spraying a specimen it is important that the nozzle of the gun should be far enough away to ensure that the blast of gas and atomised solution is not strong enough to disturb loose pieces. In some cases it is necessary to erect a polythene tent over the specimen and to direct the spray upwards at an angle to the object, so that the solution may fall upon it as a fine rain. By using such a technique even impressions in dry sand can be consolidated, provided that an emulsion of polyvinyl acetate is the consolidant and enough time and care is taken.

Spraying is about the only effective way of getting consolidation materials into the interior of large skulls. The natural openings, the foramen magnum and the nares, are the obvious points for injection.

There is a limit to the solid content of a solution which may be used in a spray gun for if it is too high and the solvent very volatile an effect known as 'cobwebbing' will occur. The solid is deposited in the air as very fine filaments which stick to each other and so form synthetic cobwebs.

When using a spray gun the preparator should wear a mask.

Application by Immersion

The total immersion of a specimen in a solution or an emulsion is sometimes the best means of consolidation. It takes considerably less time than any other method, but cannot be used safely on all specimens. A fragmented specimen should not be treated in this manner unless the pieces are large enough to be considered as separate objects, which are later to be put together to make a whole. That is, each piece must be large and strong enough to be handled without crumbling or breaking. Similarly, very friable fossils should not be treated by immersion as the force exerted by the fluid entering the pores suddenly at all points may be enough to cause the total collapse of the specimen.

When this method is used the fossils should not merely be placed into a vessel containing the solution and at the end of the treatment be pulled out either by instruments or by hand, for no strength is imparted to them by any means of consolidation until the plastic has dried. In a wet condition the fossil is much more likely to break than before it was immersed. The specimen should be supported on a flat-bottomed basket made of wire or some other suitable form of mesh and the whole lowered carefully into the liquid. The basket should have handles which stand above the surface. After a period of some hours the basket containing the fossil can be removed slowly and carefully from the bath. It must be remembered that force is exerted upon the specimen by the liquid through which it moves and the basket should not be jerked out. When out of the fluid the fossil, still in its basket, should be allowed to drain and it should then be tested by gently touching it to ascertain whether it may be safely handled. If it is considered to be too weak, another piece of mesh fashioned to fit easily into the basket, and equipped with side-pieces on which it may stand, is lowered gently until the flat top just touches the fossil. The whole is then inverted and the basket removed. The specimen is now on a mesh support which stands clear of the bench top. If the first test has shown that the fossil is strong enough, it may be taken out of the basket and placed on the bench to dry, or supported on a series of knife edges formed by bending strips of zinc or painted tin plate into a series of W's (Fig. 1).

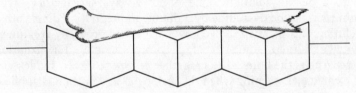

Fig. 1. Specimen drying on a metal 'zig-zag'.

The immersion method should, whenever possible, be avoided for large specimens, except where polyvinyl acetate emulsion is the consolidant, as it requires the provision of large vats of inflammable and volatile substances with the attendant risk of fire and, if certain solvents are present, an increased physiological danger. When it is unavoidable, however, special arrangements are required to cancel out these hazards.

Vacuum impregnation is used in some establishments but the author has yet to be convinced that, for general purposes, the size and expense of the apparatus required is justified by a result superior to that obtainable by other methods. This is not to say that it does not have its advantages in special instances, but most of these can be dealt with either in a vacuum desiccator or under a bell jar fitted with an L-shaped plastic gasket and standing on a metal plate equipped with legs and two vacuum taps. When vacuum impregnation is carried out it is important to remember that volatile solvents boil at low pressure, and what is often mistaken for the very efficient removal of air from the specimen under treatment, is in fact the formation of bubbles caused by this effect. Such low pressure boiling can be violent and may damage the specimens.

Special Cases

The materials and methods mentioned above can be used with success on the majority of specimens, but some require special consideration. The most common are dealt with below.

Bone with much Cancellous Tissue

Certain bones, like the human femur and the antlers of deer, have large cancellous areas and the consolidation of such specimens with a plastic solution alone is not always satisfactory. This difficulty can be overcome by a double impregnation, first with a solution of a plastic followed, when this has dried, by treatment with a diluted emulsion of polyvinyl acetate. The solution strengthens the bone, sticks together any loose pieces and lessens the chance of warping when the water-based emulsion is applied. In the majority of cases the treatment with the solution will make the bone sufficiently robust to stand immersion in the emulsion. One part of the emulsion, as it is provided by the makers, is diluted with two parts of distilled or de-ionised water. The dried, plastic-consolidated specimen is immersed in this and left for several hours. At the end of this time the preparator should place his hand into the emulsion and feel the bottom of the vessel to ascertain whether the solid particles of the emulsion have deposited or not. If they have, the emulsion can easily be reformed by agitating it with the fingers, the specimen may then be lifted out and the excess emulsion allowed to drain off. It is then placed to dry on a piece of metal folded into W's. From time to time it should be turned over and any excess emulsion wiped off with a wet rag. It is important that 'tears' of emulsion should not be allowed to dry on the surface, for if they do they can only be

Wet Bones

The treatment of wet bone depends on the localities from which it comes and its condition. When it has a plastic soapy feel, as is sometimes the case with bones coming from sandy localities, it will often become stronger if allowed to dry very slowly. This is best done by keeping it in an enclosed space and gradually reducing the humidity of the air within that space. A polythene 'tent' and small vessels of water can be used to carry out this operation or the latter may be replaced by porous pads soaked in water. The introduction of a volatile fungicide, such as thymol, into the chamber will reduce the possibility of moulds growing while the process is going on. Smaller specimens may be covered by a bell jar. When the fossil is dry enough it should be impregnated with a consolidating solution by careful spraying or the use of a pipette. In order to cancel out the effect of any water that may remain in the specimen, a small amount of a high boiling solvent, such as n-butyl acetate or diacetone alcohol, should be included in the solution.

Bones excavated from beaches are almost certain to contain some salt and this must be washed out. They should therefore be kept wet until they arrive at the laboratory and there washed either in running tap water or constant changes of still water until the washing water shows no more salt content, when tested with silver nitrate solution, than is normally found in mains water. The bones may now be taken from the water and allowed to drain but not to dry. They are then immersed in the 1:2 dilution of polyvinyl acetate emulsion in water and then dried as described above. Wet bones containing no salt may, after the removal of all mud, be treated with emulsion immediately. Dr Purves first used polyvinyl acetate emulsion in this way to treat the animal bones excavated at Star Carr.

An alternative method for dealing with wet bones is to dehydrate them in n-butyl alcohol. This alcohol will displace water from a porous substance in a very spectacular way. When a wet specimen is suspended in it, the water will pour out in a visible stream and collect at the bottom of the vessel.

Consolidation and Repair of Specimens

A large jar is needed and some means of suspending the specimen must be provided. This can be a sheet of thin aluminium with holes punched in it and with strips which will hook over the edge of the jar. The level of the basket should be well above the bottom of the jar. Since n-butyl alcohol has an irritating odour a cover of plastic sheet or aluminium foil should be fitted over the mouth of the vessel. The jar is filled to the required level with n-butyl alcohol and the specimen on the support lowered into it. The fossil should be left there for some hours and when no more water is seen to leave the specimen, it can be taken out and consolidated with a solution of a plastic which is soluble in n-butyl alcohol such as Butvar B98 or other polyvinyl butyrals.

Wet Wood

This may be placed into molten polyethylene glycol 4000 after the excess water has been drained off. The wax melts at about 50°C so there is no fear of damage due to the water within the specimen boiling. The length of time that the specimen needs to be in the wax bath depends on its size, but at least a day is required to be sure of full penetration into a small specimen. After treatment, as much of the excess wax as possible should be removed while it is still molten; that which remains may be removed by local reheating or washing carefully with swabs soaked in hot water. Wet shale and jet have been successfully treated in this way. Alternatively the specimen may be immersed in a 50–60 per cent solution of the wax in water and the water slowly allowed to evaporate. Before the solution becomes too viscous the specimen is removed, the excess solution wiped off and the object allowed to dry in air. This process takes several weeks.

A low voltage heating iron which has been found of great use in working polyethylene glycol is described in Appendix I.

Elephant Tusks and Teeth

Recent elephant ivory is one of the strongest and most elastic of natural materials. Fossil tusks from some localities remain in this happy condition and require no treatment other than storage in an atmosphere which is neither too wet nor too dry.

Almost the whole of an elephant tusk is composed of dentine, but this differs from that found in ordinary teeth. The tubules within it are not straight and as a result of this, when the material has lost most of its organic content and dries, it warps in two directions and the inner part of the tusk shrinks from the outer layer and breaks up into widely separated, roughly cubic pieces. The outer layer is of a different cellular structure and does not break in this way, but owing to shrinkage it often cracks longitudinally and may warp and so produce wide fissures. A tusk in this condition is one of the most difficult fossils to conserve.

A dry tusk in this state may be treated in much the same way as has been described for cancellous bone. The outer layer should first be treated with several applications of a solution of a suitable plastic (Butvar, etc.). Where there are cracks the same solution should be pumped through them into the inside, with a pipette. The plastic should be allowed to dry and then a further application to the inside of the specimen should be made through the open proximal end of the tusk. In cases where the previous treatment has made the specimen strong enough to be handled it can be placed so that the solution may be poured in; otherwise a spray is required. The tusk should now be left for several days to dry and then, if there are cracks in the outer layer, the specimen should be bound at intervals with string as tightly as possible without causing breakage. The object is to close the cracks as much as possible and to guard against any undesirable warping in the next stage of treatment, which is impregnation with polyvinyl acetate emulsion. In the case of specimens up to 1.25 m in length this may be done by immersion in a 1:2 dilution of the emulsion in deionised water. Where there is no vessel available to take a specimen of this size, one may be improvised by draping heavy duty polythene sheeting into a wooden box. The specimen is placed on a support which may be made of slats of wood and lowered into the liquid. It should remain there for 24 hours and is then removed on its support and the excess emulsion allowed to drain off. While it is still wet the specimen should be tested to establish whether it is possible to reduce further the width of any cracks by applying tighter binding. Should this be the case, the new bindings may be of thick copper wire, the ends of which are twisted together in order to apply the required pressure. When the wire is in position the string can be removed. The tusk should now be placed on a clean support and left to dry in air and should be examined from time to time and any 'tears' of emulsion which form on the surface wiped off with a wet rag. When it is almost dry, the bindings may be removed and the emulsion, which will have collected round them, wiped away. One immersion is usually enough but, if it is thought that the specimen would benefit from further treatment, it may be repeated provided that the first application of emulsion is completely dry and solid.

The final stage of treatment is carried out after the tusk has been allowed to dry for some days; then undiluted emulsion is painted over the whole outer surface and is forced into any cracks with the brush and the excess is wiped off with a wet rag. Emulsion may also be poured into the open end of the tusk. Any cracks which have not been filled with emulsion may be packed with a jute floc based gap-filling cement or plaster of Paris.

Should the specimen be too large for immersion, the first part of the treatment with the plastic solution is the same as above but the emulsion will have to be applied by brush, pipette and by pouring and will take a very long time to complete.

Tusks which are wet when collected should, whenever possible, be kept so until they arrive at the laboratory, where after any mud has been removed and it has been established that they contain little or no salt, they may be immersed in emulsion without prior treatment. Salt when present in quantities in excess of that found in the normal water supply should be washed out before the specimen is placed in emulsion, because if this is not done, the solid particles of the plastic will be precipitated. It is not possible to wash very large tusks. The best way to treat those containing salt is to encase them completely in a thick coat of paper pulp or surgical cellulose soaked in de-ionised water and to allow the whole to dry slowly. The salt will be drawn out of the specimen and deposited on the outside of the casing. The slow drying should prevent the tusk from disintegrating, and it may then be treated in the normal way.

Elephant molar teeth are somewhat easier to deal with. They are formed of dentine plates covered with enamel and held together with cementum. The plates are normally robust but the cementum is liable to fracture or to break up through desiccation. The pieces of a badly broken tooth should be consolidated separately and then the whole repaired with one of the quick drying cements mentioned above. Where cementum is missing it can be made up with a jute floc gap-filling cement. Double impregnation, using first a plastic in solution and then polyvinyl acetate emulsion has proved to be the best treatment for these specimens so far developed.

All the plastics used in consolidation are good adhesives and if specimens are placed while wet on a bench and left they are very likely to stick to it. This could be disastrous and can be avoided easily. Bench tops which are used when consolidating specimens should always be covered with some sort of paper. Cheap lining papers used by decorators do very well. More expensive papers coated with polythene on one side, may be obtained from most suppliers of chemical apparatus. Specimens will not stick to the surface on which they are resting if they are continually moved but if they do, by some mischance, get stuck to a paper surface the area of paper involved can be cut out of the main sheet and detached from the fossil by using a suitable solvent.

The strips of metal bent into a series of W's, which have been mentioned before, are another way of avoiding the unwanted adhesion of a fossil to a fixed surface. Once more, if the specimen is moved occasionally it will not stick at all, but if it does in this case the areas concerned are very small and the object can be freed with ease.

The solutions and emulsions will become dirty in time, or the lids will be left off the containers and the contents will dry out. The time required to clean out jars is wasted so it is therefore better to use containers which may be thrown away when they become fouled. There are many products which are sold in glass jars and these when thoroughly washed, are excellent for storing consolidating solutions. Where several people work in a laboratory a stock of these useful and expendable articles can soon be built up. It is important, however, to remove all original labels from them and re-label them clearly with their present contents. Bottles which have contained anything drinkable must never be used for this purpose.

Consolidation and Repair of Specimens
Procedure in the Repair of Fossils

There are a number of points of procedure in the repair of fossils which are of prime importance. It has been said before, but bears repeating, that friable or comminuted specimens must be consolidated before any attempt is made to stick the pieces together. Before adhesive is applied each joint should be made 'dry' in order to see that it is as tight as possible over its entire length. Bad contact is usually caused by some foreign matter on one or both contact surfaces and this must be removed. It may be necessary to make several 'dummy' runs before all such obstructions can be located. Normally only two pieces should be joined at a time and care should be taken to see that no overhang occurs which may lock out another piece which has yet to be found. That is to say, no joint should be made which forms an open space bounded on all sides, or that a space that is broader on the inside than it is on the outside is produced (Fig. 2).

In order that a joint should not be distorted the two contact edges must be held so that the accuracy of the fit is maintained until the adhesive has dried. Fossils are of irregular shape and, are in most cases, delicate objects; so it is only rarely that clamps may be

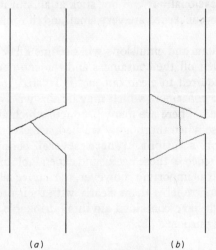

Fig. 2. Diagrammatic representation of a partially repaired specimen. The gap in (a) is acceptable, that in (b) is not.

used to hold two sections of a joint together. Occasionally rubber bands can be arranged to maintain the contact between two small robust pieces. The normal method of support is to bury one piece, with the contact edge uppermost, in sand contained in a tray or box, and to place the other contact surface upon it. By moving the section buried in the sand a point of balance can usually be found so the two parts of the joint will remain in accurate register. In some cases the length and distribution of weight in the upper piece makes it impossible to achieve a true balance, so a point nearest to equilibrium is found and the top section is held in position by clamping it lightly in a retort stand. Should it be too heavy to be supported in this way a framework of laboratory scaffolding is built around it to hold it in position. Laboratory scaffolding is very versatile and may be obtained from most suppliers of chemical apparatus. Illustrations of the various pieces available are to be found in the catalogues of these firms, so making it easy to select those most likely to be of use. It is obvious that the size and depth of the sand tray required is directly proportional to the size and weight of the specimen. A range from about 5 cm square and 1.5 cm deep, to 2 m square and 30 cm deep will cope with most specimens.

Sauropod limb bones can be enormous in size and very heavy. A scaffolding of some form of perforated angle iron, such as 'Dexion' or 'Handy Angle' can be used with advantage in these cases, for should the bone fall in the last stages of repair it will not only be damaged but someone could be badly injured. Guy ropes fastened about the bones and tied off to strong eyelets fixed into the sides of the sand tray have been used to hold such bones in position.

After the 'dry' joint has been made and supported satisfactorily, an appropriate adhesive is applied to both contact faces and they are pressed together. The joint is now examined visually and felt with the fingers to ensure that it is tightly and accurately made. Any excess adhesive is cleaned off with a soft brush moistened with the appropriate solvent and the specimen is left until the adhesive has dried.

The choice of adhesive depends on the size and weight of the fragments. For light pieces, even if they will eventually build up into a large and heavy specimen, it is better to use one, such as

28 Consolidation and Repair of Specimens

Butvar B76 in acetone, which dries rapidly, thus saving time. As the fragments of a large specimen are reassembled the sections become larger and heavier and at some point the adhesive must be changed for one which is strong enough to support them. For those which are about 30 cm long and have a broad, tightly fitting contact surface, polyvinyl acetate emulsion may be used. Although this is an excellent adhesive, it remains elastic and is not suitable for making joints in the middle of long narrow specimens, such as the mandibles of the long snouted crocodiles. Joints of this nature are best made with one of the jute-floc based gap-filling cements (Fibrenyl and A.J.K. dough) which are also effective for joining sections of very long and heavy specimens including the limb bones of giant sauropods. When using one of these adhesives it should be mixed with enough of the basic plastic solution to give a lump-free paste which can be pressed out of the joint. The paste should be applied to both contact faces and with one part of the specimen seated in the sand tray the other should be pressed on to it. If the specimen is large and strong enough, tapping the top piece with a rubber hammer helps to achieve a tight joint. Excess adhesive is collected on a steel spatula and used to build up any areas which are missing; then the whole joint is cleaned up with a brush soaked in solvent.

Dowelling may be advisable in some very heavy specimens. It is important that the dowel holes should be as near to the centre of the bone as possible, they should be exactly opposite each other, and should also be wide enough for there to be some play between the sides and the dowel rod. A hole of the right diameter should be drilled into one piece; the other piece with the joint face up is then placed in the sand tray. The edge of the hole is now circumscribed with thick paint and the piece containing it is lowered as accurately as possible to fit that in the sand tray. The two are pressed together and the top one lifted off. The paint will mark the point at which the second hole must be drilled. The iron dowel rod should be cut so that its length is a little less than the combined depth of the holes into which it is to fit. The two pieces to be joined are fitted together with the iron in place and the joint edge examined to see that the rod has not distorted the fit. If it has, one or both of the holes must be enlarged to correct the error. When this has been done the hole in the section in the sand tray is filled with a gap-filling adhesive, the rod pushed down into it and the rest of the joint surface covered with adhesive. Adhesive is then applied to the other piece and the hole in this almost filled. The two pieces are now joined and pressed or hammered together with a rubber mallet. Except when made with epoxy resins, large joints take several days to harden, and should be left undisturbed for at least a week.

Variations on the dowelling technique can be applied to some elephant tusks. Occasionally the specimen is robust but broken transversely into pieces about 30 cm in length and the centres are soft enough for a hole to be bored through each segment. The drilling may be done with augers or masonry drills fitted with extension pieces. It is safer to use a hand operated brace rather than an electric drill, unless the latter can be regulated to a very low speed. Holes are bored in the joint faces of the tip and the proximal end pieces of the tusk to approximately half their depth. The other pieces are drilled right through their centres. The holes may be enlarged with a rasp.

An iron bar of rectangular cross section is selected of a size which will allow two pieces to fit into each hole so that they overlap and there is a space between them and the sides of the hole. The bar is cut into pieces long enough to pass through each hole in the sections of tusk and leave about 10 cm projecting from both ends. The proximal end of the tusk is placed, with the drilled joint surface upwards, in a sand tray. Some jute floc gap-filling cement is mixed and the hole filled with it then a length of iron which will leave 10 cm projecting is then pushed into the hole so that it is off centre. The iron which is to pass through the next section of the tusk is pushed in beside it, the joint surfaces of the piece of tusk in the tray and that of the one which fits onto it are spread with the cement, and some is pushed into the hole at the lower end of the upper section. This is then threaded over the projecting iron and the joint pressed together. The next piece to be joined should now be placed into position, without any cement, to make sure that the projecting iron will allow the second joint to be made accurately. When the first joint has set the loose section is removed and gap-filling cement is pushed in to fill the hole in the piece which has been joined to the proximal end and the iron bar for the next section is pushed into it to a depth of 10 cm. Cement is spread on the two joint faces and the next section of tusk is threaded over the bar and the joint made. This procedure is repeated until the whole tusk has been assembled. Figure 3 illustrates this method of overlapping dowels.

Very long straight tusks have been found in some pleistocene gravels and have been successfully extracted intact. However their condition, weight and length makes it very easy for them to be broken when they are moved in storage. The outer surface, more often than not, has been broken up during fossilisation and recemented by iron bound sand and is indistinguishable from the inner parts of the fossil. In such a case nothing is lost if a channel is cut in the worst side of the specimen, an iron bar inserted and the channel filled with jute floc gap-filling cement or plaster of

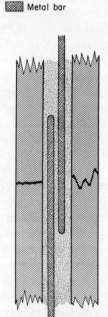

Fig. 3. Section through a joint with overlapping dowels.

Paris. The channel should run from about 15 cm from the tip of the tusk right out to the proximal end. The specimen should be checked to see that it has been efficiently consolidated and then supported on the underside by packing before the operation is started. The course of the channel is marked out with chalk and it is cut using dental rotary saws and small sharp cold chisels. The chisels should be used by hand as far as possible. If a hammer must be used great care is needed.

The channel should be twice as wide as the bar which is to be placed into it, and be half the depth of the tusk. When it has been cut it should be treated with a plastic consolidating agent. The iron bar is cut to length and bent to fit the channel and, if plaster of Paris is to be used as the filler, the bar should be painted with epoxy varnish or polyurethane paint to prevent rusting. Some of the filling material is placed into the channel. In the case of plaster of Paris this should be mixed so that it is spreadable and does not flow easily. The bar is lowered into the cavity and then covered with the filling which should be pressed in with a spatula in order to avoid the formation of air spaces. The filling is built up to the edges of the channel, and smoothed off. When dry it is painted to resemble the tusk. It is obvious that this operation requires a great deal of skill and takes a great deal of time. It should therefore only be done if it is absolutely necessary.

The techniques described above are valid for the repair of a wide range of fossils, and are easily applied to specimens which have been broken into a few pieces, because the correct contacts between the fragments can be found very quickly. However, when one is confronted with a mammalian tooth row which has been shattered into hundreds of pieces, or an entire skeleton which has suffered the same fate, the problem is quite different. At first sight teeth in this condition may seem to be beyond recovery, but provided that there are some fairly large pieces and the rest can be handled with forceps, something of value can usually be saved. Vacuum tweezers which are invaluable for this type of work, or for sorting small fossils, are described at the end of this chapter.

The best approach is to sort out all the more obviously associated fragments and group them together in small trays. The cusps and ridges of the crown are often recognisable and together with other pieces of enamel form one logical group. In the same way pieces of root are identifiable and these should be isolated. Any doubtful pieces are best sorted according to shape and colour.

The search for fits should start with the pieces of crown. At first this can only be done on the same principle used to assemble a jigsaw puzzle. That is, the broken edges are compared and when the contours of one are found to fit into those of another it may be assumed that they join. Each supposed fit should be examined under a binocular microscope, for the outlines of broken edges of small pieces often appear to the naked eye to be a true junction, but magnification will reveal that this is not so. When the fit has been established the two pieces are joined with a rapidly setting adhesive. A solution of polymethylmethacrylate in ethyl acetate is excellent for this type of work. These operations are repeated until enough pieces have been assembled to make it possible to say which tooth or teeth are represented; then by comparing the reunited fragments with complete teeth of a similar specimen or, in their absence, illustrations in books, the finding of the missing pieces is made easier. In the same way the roots are reassembled and united to the crowns. After all the pieces have been joined, any parts of the teeth that are missing may be restored in plaster of Paris or one of the polymethylmethacrylate monomer/polymer mixes. The former can be carved easily when set and the latter can be modelled with a brush moistened with ethyl acetate before it has set hard.

A general knowledge of anatomy is essential when repairing any vertebrate fossil, particularly if the fragments of several bones have for some reason or another, become mixed. Without it much time is wasted. The sutures and foramina which occur in skulls are valuable clues as to which pieces are likely to fit together. When the whole or part of a skeleton has to be dealt with, the distal and proximal ends of the limb bones are usually in a recognisable condition and can be separated into lefts and rights. Parts of the shafts of long bone can occasionally be recognised by their thickness and shape. The vertebral centra are often robust and can be separated from the rest of the bones. In the same way, parts of the neural arches can be recognised and collected together. Heads and fragments of shafts of ribs are distinguishable and these too can often be separated into lefts and rights. Some parts of the pectoral and pelvic girdles have distinguishing features which are fairly obvious. It follows that if as many of the fragments as possible can be grouped by anatomical features, fits between them will be easier to find, however there will always be pieces which cannot be assigned on anatomical grounds to a particular part of the skeleton and these have to be fitted by trial and error. Useful clues are similarities in colour and thickness. The fragments are joined with an adhesive which is suitable to the size and weight of the specimen, using the techniques previously described.

The Repair of the Surfaces of Bones Damaged by the Shrinkage of Animal Glue

In the section dealing with materials used in the past to consolidate specimens it was said that animal glue, due to shrinkage over a long time, could contract and damage the surface of a delicate bone. It is sometimes possible to repair this damage, but it takes time and the treatment is not always successful.

Small areas of the glue are softened with hot water applied with a small brush, and when it has become malleable the fragments of bone attached to it are treated on the undersides with polyvinyl acetate emulsion and gently eased into their correct position. It may be necessary to apply pressure until the emulsion has solidified, and this may involve the exercise of some ingenuity. The main problem is to avoid the pressure device sticking to the glue. This may be done by covering the contact surface with either paraffin waxed paper or a small piece of silicone treated paper. When the first area is solid the treatment is continued in stages until the whole surface has been restored to its normal condition. After all the polyvinyl acetate emulsion has dried out, the remaining surface glue is softened in small areas and carefully scraped off. The last stages of the removal should be done with a soft brush. When all, or as much as possible, of the glue has been removed the specimen may be impregnated with a dilution of polyvinyl acetate in water in a proportion of 1:3 v/v.

Vacuum Tweezers

Those commercially available are intended for use in instrument making and light industry. They allow a small object to be picked up and moved into position with the application of minimum force, and are extremely useful in any palaeontological techniques in which small fragments or fossils have to be handled. The apparatus consists of a small electically driven vacuum pump to which is connected a length of flexible tube, which in its turn is fitted to a metal handpiece about the same length as a standard ball point pen. At about 2.5 cm from the forward end of the handpiece there is a hole approximately 0.5 cm in diameter. The free end of the metal handpiece is fitted with a nipple made to fit the hole in the base of a standard hypodermic needle. The needles used are about 8.0 cm long and are supplied in a variety of internal diameters. At about 2.5 cm from the point the needle is bent at an angle. The point is not sharp, but is cut off flush. Rubber suction discs of various sizes may be bought to fit the ends of the needles, but these are not much use for handling fossils. It has been found that the performance of this apparatus is improved if about 1.5 cm of polythene capillary tube is pushed onto the needle leaving about 2 mm extending beyond the end.

In use the pump is switched on and the end of the polythene tube is placed on the fossil to be lifted. The tip of the forefinger is then pressed over the hole in the handpiece, thereby creating a vacuum which will hold the specimen onto the end of the tube. The specimen can now be lifted and placed into position and, when this has been done satisfactorily, the finger is lifted from the hole so breaking the vacuum and releasing the object.

In Appendix I there is a description of the method employed to heat polythene tubing so that it may be drawn out to make a pipette. Plastic capillary tubing can be made in the same way.

E. H. Stinemeyer (1965) described a similar piece of apparatus which he called a vacuum pick. He made this specifically for removing microfossils from disaggregated samples (sic) and details the type of aquarium aeration pump he modified, the method for making a handpiece fitted with a 200 mesh filter backed by cotton wool (cotton) and the hypodermic needles he found to be most useful. The paper is well illustrated with technical drawings and by following his instructions the apparatus could be made easily by preparators living in countries in which the commercial apparatus is not available. Fitting a filter into a standard piece of apparatus is not difficult and prevents small particles or fossils from being sucked into the pump.

References

Bronson, Christensen B. (1971) *Conservation of waterlogged wood in the National Museum*, Copenhagen: National Museum Press.

Brink, A. S. (1956) 'On the uses of "Glyptal" in palaeontology', *Palaeont, Afric.*, IV, 124–30.

Cornwall, I. W. (1956) *Bones for the archaeologist*, London, Phoenix House Publications.

Hempel, K. F. B. (1969) 'The restoration of two marble statues by Antonio Corradini', *Studies in Conservation*, **14**, 126–31.

Larney, J. (1971) 'Ceramic restoration in the Victoria and Albert Museum', *Studies in Conservation*, **16**, 69–82.

Newman, B. (1955) 'The use of fibrenyl', *Mus. Journ.*, **55**, 65–6.

Shorer, P. (1967) *Introduction to the products of Quentplass Ltd*, U.K.G.I.I.C. Lecture.

Stinemeyer, E. H. (1965) 'Microfossil vacuum needle segregating pick', *Handbook of Palaeontological Techniques*, ed. B. Kummel and D. Raup, San Francisco: W. H. Freeman & Co.

Werner, A. E. (1958) 'Technical notes on a new material in conservation', *Chron. d'Egypte*, xxviii, 273–8.

3
Collecting and Treatment in the Field

As every professional palaeontologist knows and fortunately the bulk of amateurs too, the haphazard collecting of fossils is a thing to be deplored. Unless the exact locality and stratigraphic position of a specimen is known and recorded it has little or no scientific worth. For this reason the most important pieces of field equipment are a note book, a pencil and some suitably sized pieces of strong paper which can be used to label the fossils as soon as they have been dug out. A more detailed discussion of labelling will be found at the end of this chapter. A measuring tape, a prismatic compass, a clinometer and a map are desirable, especially when a systematic collection of invertebrates is being made from an exposed vertical section. A description of stratigraphic methods as applied to collecting is not within the scope of this book, but a plea is made to any budding collector who reads it to study these methods either from books, or better still to learn them from a person experienced in the scientific collection of fossils.

Tools for Field Use

The tools used in collecting vary with the type of rock from which the collection is to be made and depend very much on the personal choice of the collector. Except for specimens which may be found loose on a surface, such as a beach, all fossils have to be dug out of the containing matrix in some way and as has already been said their exact locality must be recorded and they must be labelled in the field.

For the actual removal of the fossils the tools range from fine dental scrapers to pick axes. In localities in which the matrix is soft a small pointed trowel is a most useful implement. Some sort of pick is nearly always desirable and a 'Hardy' pick, which can also be used as a hammer, is a very convenient type. The head is attached in such a way that it can easily be removed, but, short of breaking the shaft, cannot become detached in use. An ex-army entrenching tool may be employed to clear the face of a section and has many other uses.

Nowadays, geological hammers may be obtained in designs which extend from those differing very little from the first types made for this purpose to others of various shapes and sizes in which modern methods of tool making and materials for the shafts are employed. Almost everybody has his own opinion as to which is the best pattern. The present writer prefers to use a piton hammer made for mountaineers, but, as far as is known, is alone in this choice.

In hard matrices, such as limestones and sandstones, cold chisels are needed. These should be stout enough in the shaft to withstand rough usage and long enough to allow the cutting edge to be driven two or three inches into the rock and still have enough shaft protruding for easy handling. Blade widths, for most uses, can be from $\frac{1}{4}''$ to $1''$ (0.5 to 2.5 cm) although for splitting slates and shales wider blades may be an advantage. Chisels are cutting tools and should not be used as levers or they are likely to break. A small crowbar should be carried for this purpose. Although a geological hammer can be used with cold chisels it is by no means ideal for this purpose as its shaft is normally too long and the edge can flake dangerously. The type of two-faced club hammer, made for bricklayers, is much easier to use. A two pound (approx. 1 kg) hammer of this type will cope with most situations.

Fossils are sometimes cracked, comminuted or friable and need to be consolidated before they can be excavated or must be treated as soon as they have been dug out. It is therefore necessary to take into the field consolidating agents and an adhesive for sticking back into position detached pieces. What is taken will depend upon what is available, but a good selection would be one container of a thick solution of Butvar B76, or a similar polyvinyl butyral, in acetone to be used as an adhesive, another of a thinner solution of the same plastic in the same solvent for consolidation, and a third containing an emulsion of polyvinyl acetate diluted with distilled or deionised water. The containers should be unbreakable and be equipped with tight-fitting leak-proof closures.

Because on all types of excavation there are always loose particles of matrix about, it is better not to use these fluids by dipping brushes into the bulk container as they will very quickly become fouled. Before use, as much as is needed should be poured from the stock container into a wide-mouthed vessel. Jars of this type can be made easily by cutting the tops off the plastic containers in which household washing-up liquids and other detergents are sold. These have the advantage of being free, light and unbreakable.

A selection of small brushes for the application of the fluids are required and a wide flat brush such as a painter's 4" (10.0 cm) dusting brush is useful for cleaning sections and specimens.

Newspaper and tissue paper and a supply of small cardboard boxes and collecting bags will be needed to pack the specimens and a reel of 'Sellotape' or some other transparent sealing tape is useful to secure the packages.

Certain items for personal comfort and safety should be included in the field kit in some circumstances. When a hammer and chisel are being used on hard rock, chips can fly back towards the collector with considerable force and can damage the eyes. This is a hazard which should not be overlooked either in the field or in the laboratory. It can be avoided if industrial goggles are worn. These are both light and cheap and some are designed to be worn over spectacles. By using them painful and dangerous injuries to the eyes can be avoided.

Less seriously, during the same operations, the hammer may miss the chisel and hit the hand. Pain and possible injury, may be avoided if a leather glove is worn on the hand holding the chisel. When working below cliffs or in steep-walled quarries, a caving or industrial helmet should be worn as even a small stone falling from a height can cause serious and possibly fatal injuries.

Consolidation in the Field

The consolidation of specimens in the field should only be carried out when it is absolutely necessary and there is some chance of the consolidant drying and strengthening the specimen. Fossils which are strong enough to be handled and show no signs of collapsing are best left untreated. Careful packing is all that is required to ensure their safe arrival back at the laboratory. Old newspaper is a very convenient packing material; if torn into the right sized pieces and crumpled up, it makes a very resilient packing bed. In the flat state it may be used for simple wrapping. Owing to its porosity it will hold much water, and can be used when it is necessary to keep a specimen wet until it can receive proper attention. Tissue paper is useful for packing more delicate fossils. Cotton wool, known in America simply as cotton, is not recommended as a packing material for highly ornamented specimens or small teeth, as the filaments can be difficult to remove from the specimen and the fossil may be damaged when the wool is taken off.

If the specimens are dry and in a condition that makes consolidation necessary there is really no problem as one of the plastics in solution may be used. It is better to apply it and allow it to dry before the specimen is excavated.

When a fossil is damp but not waterlogged and it is thought that after it is removed from the earth it will dry out in the time at your disposal, it should be excavated with a fair margin of matrix attached, labelled and set aside. It may be treated immediately with diluted polyvinyl acetate emulsion, or if there is no fear of it collapsing on drying, it should be left to dry and then treated with a polyvinyl butyral solution. Butvar B76 has been mentioned in the collecting kit because it is dissolved in acetone and will therefore dry rapidly, but Butvar B98 in isopropyl alcohol is also suitable.

Very wet specimens present rather a difficult problem. The wetness may be temporary, due to seasonal rains, and if this is the case it is best, when possible, to defer the collecting until the locality has dried out. If the wetness is a permanent feature, the action taken is largely dictated by the time and resources available.

Where time is short there are three possible approaches. The first is to remove the fossils from the beds with as much matrix surrounding them as is practical. No more should be cleaned off the specimen than is enough to see that it is there and worth having. The whole should be placed in polythene bags and the mouths sealed with plastic adhesive tape. The bags should be packed in rigid boxes on a bed of crumpled newspaper. More crumpled newspaper is placed around and above the bagged specimen to stop movement in the box, a label is put inside and the box closed. Other labels should be placed on the outside of the box. This method has been used by many collectors with success.

Another possibility has not been fully tested but is worth mentioning. That is to use 'Quentglaze' to consolidate the specimen. Should this be done it must be remembered that this plastic is very resistant to common solvents and should therefore be used sparingly. As was mentioned in the description of 'Quentglaze', it has been used with great success to consolidate the wet sand surrounding a wet specimen and this acts as a support while lifting the fossil. In these cases the plastic was not applied to the specimen.

Provided that the expedition can stay on site long enough, a shelter of tarpaulin or heavy duty polythene sheet can be erected and the specimens either allowed to dry under this before treatment with a solution of plastics or they may be cleaned of matrix and treated on site with polyvinyl acetate emulsion while still wet and allowed to dry under the shelter. If Butvar, or any alternative, dissolved in an organic solvent is the consolidant the hazards presented must be remembered; they should not be used in badly ventilated shelters. Considerable build up of vapours can occur when specimens are dried in enclosed space. Any shelter built for this purpose must be opened and air allowed to flow freely through for sometime before anyone enters to work on the collection. Remember that polythene sheet is inflammable.

In some localities the beds contain so many fossils that the isolation of them in the field, except for any particularly choice ones that can be seen, is best not attempted. This type of material should be collected in blocks of matrix and the removal of fossils postponed until the samples arrive at the laboratory where many more are likely to be extracted in good condition than could be removed by scratching at the beds in the field. This particularly applies to some cave breccias, bone beds and fissure fillings. Good packing and labelling, and sometimes a small amount of consolidation, is all that is required. In caves, unless they are very well ventilated, it is unwise to use any consolidating solutions in volatile solvents.

The preceding remarks obviously apply only to small or medium-sized fossils of any type which can be collected by one or two persons on a short excursion. The excavation of large vertebrate remains presents other problems.

The Collection of Large Vertebrate Specimens

Vertebrate remains are found in all types of matrix from soft clays and gravels to hard limestones and sandstones. Whatever the matrix, they can never be extracted successfully in a hurry. If time is not immediately available to do the job properly, except in cases of 'rescue' operations, it is better to wait before doing anything at all.

Often vertebrate remains are uncovered during excavations made for commercial reasons. Owners of sites and contractors are nearly always as accommodating and as helpful as they can be, but sometimes for economic reasons, they cannot, nor can they be expected to, hold up operations for any length of time. In such cases, collecting should be selective, and skulls and limb girdles given priority. If it is at all possible to photograph all that is exposed of the animal before work starts and any parts which are subsequently revealed, this should be done. In these conditions it is better to extract one or two good specimens, than to wreck everything by trying to do more than is possible in the time available.

When time is no object, the methods used to collect the specimen or such parts of it as are preserved, will partly depend on the matrix. Whatever the type of rock there are some things which should always be done.

The locality of the find and its stratigraphic position must be recorded accurately. A map reference, as well as the name of the place, should be given. The area surrounding the specimen should be searched for loose pieces of bone and, when there is an obvious fit between any of them and the main specimen, they should be stuck back in place. Those for which fits cannot be found should be strengthened with a consolidating solution, and when dry, packed with a label saying where they were found in relation to the main specimen. Any exposed bone should be strengthened when necessary and a large sketch made showing all of them. A polaroid camera is particularly useful for recording such information.

Each bone is then numbered and the same number recorded on the sketch. The numbering can be done with either white or black drawing ink, according to the colour of the bone. At this point, in a climate as unpredictable as that of the British Isles, it is as well to erect a shelter over the specimen. When this is not possible, large polythene sheets should be obtained for covering the specimen to protect it when necessary.

When a large fossil is embedded in hard rock it cannot be extracted bone by bone in the field but has to be lifted encased in a block of matrix. Beyond what is necessary to make sure the block contains all the specimen that is there, no surface development should be done. The general process of excavation is well known to all palaeontologists, although there are variations in detail practised by almost everybody who collects such specimens, and circumstances often dictate what must be done.

To be sure that the block contains all the bones below the surface, the surrounding rock is dug away to a depth slightly more than that of the estimated extent of the specimen downward. This leaves the specimen standing above its surroundings on a pedestal of rock. Just above the base of the pedestal a groove about 2–3 cm deep should be cut.

For the further protection of exposed bone and to give added strength, the usual method is to encase the block in plaster bandages. These are not applied directly to the specimen, as this would make their subsequent removal without damaging the fossil very difficult. The surfaces of the block are, therefore, first covered with wet paper. The type of paper does not matter provided that it will easily absorb water. Newspaper, strong tissue or cheap toilet paper are most often used. When applying the paper, particular attention should be given to narrow spaces between bones, undercuts and deep hollows. These should be tightly packed with the wet material to stop the formation of clots of tough plaster and bandage in them.

The basis of the bandages can be any strong open woven material which will absorb the plaster. Plasterer's scrim is about the best, since it is strong enough to do an efficient job and is easier to remove than other similar but more tightly-woven materials. The scrim is cut into lengths sufficient to enclose the block. The width of the strips can be anything from about 7 cm up

Large vertebrate specimens 43

to 15 cm. Lengths of more than 1 m are easier to handle if they are loosely rolled up before impregnation with the plaster.

Plaster of Paris is mixed by sprinkling into water until about 3 mm of the liquid stands above the level of the settled powder. It is then mixed into a smooth paste. The mixing is best done in a plastic bucket as these are easy to clean.

A bandage is saturated with plaster mixture and applied on top of the wet paper and is smoothed tightly down by hand. Another is taken and treated in the same way; it should slightly overlap the first along its whole length. The ends of the bandages are pushed into the groove in the base of the pedestal. This operation is repeated until one length of the specimen is completely covered and then, in the same way, another layer of bandages is applied at right-angles to the first. Depending on the size and condition of the specimen, other layers may be needed, but care should be taken not to overdo this.

When it is thought that the package needs splints for support these are best made from 2.5 cm (1") square wooden battens. A supply of this material in 2 m lengths and a small saw are good things to have on any site where bones or other fossils may need this type of support. It is, however, often impractical to carry battens, so splints are then cut from trees. In surgery and first aid, a splint is never applied to an unpadded limb and the same principle is followed in dealing with fossils. Splints should either be placed between two layers of bandages or fixed to the outside of the package by a series of tightly applied single plaster bandages.

The packaging being complete, any data which should be with the specimen can be inscribed into the plaster and then it should be left alone until the heat generated by the setting of the plaster has dissipated and the surface is cold.

The block now has to be removed from the ground. In rocks which are well laminated and strong enough and the block is not too wide, it may be possible to free it by driving stout chisels or crowbars into the bedding plane. If this cannot be done the rock below the bandage level will have to be undercut until the block is free. When this is done precautions should be taken against a sudden fall. The placing of props and wedges can help.

The block is now turned over. If something flat, like a board, can be placed under it, this makes the job easier. As it is turned, it

Plate 1. Aveley Elephant (1): Right pes, partially excavated. (By Permission of the Trustees of the British Museum (Natural History).)

Large vertebrate specimens 45

Plate 2. Aveley Elephant (2): Right pes, plaster bandaged and splinted ready for transportation. (By permission of the Trustees of the British Museum (Natural History).)

Plate 3. Fletton Plesiosaur. An example of an excavation in clays. (By courtesy of Imitor Ltd. Copyright Imitor Ltd.)

should be steadied as it passes the vertical and lowered slowly. Care should be taken to see that nobody's hands are trapped between it and the ground. Wooden battens can be placed to guard against this.

The underside is now bandaged in the same way as the upper, the ends of the bandages being carried well up the sides of the block. Where the rock is strong and the distance it has to be transported is not great and is over level ground, the bandaging of the underside may not be necessary.

The collection of large fossils from soft deposits, such as clays is much easier. In these conditions the bones can usually be consolidated and lifted separately. The tools required are pickaxes, shovels, small trowels, knives, dental scrapers, small and large brushes and sieves. The best procedure here is to clear the bulk of the soil from the whole skeleton numbering each part as it is revealed and strengthening the bones, when necessary.

When the whole animal has been uncovered, photographs, which include a scale, should be taken of the entire specimen and separate shots made of the more important parts, such as the skull, limb bones and girdles. Photographs of manus and pes are a great help when it comes to reassembling the skeleton in the laboratory. A large sketch plan showing the position of each bone should also be drawn. Larger limb bones and skulls will need plaster jackets. The procedure is to remove the matrix from the bone so that its entire length is exposed and it is uncovered to slightly more than half its depth (Fig. 4a). Consolidation is carried out as and when necessary. After the consolidants have dried, wet paper and plaster bandages are applied as already described, care being taken to tuck the ends of the bandages as far under the side of the specimen as possible (Fig. 4b). Since over half the diameter of the specimen has been uncovered the space between the ends of the bandages will be less than the width of the bone, and when the plaster has set, it will be impossible for the fossil to fall out of the casing when it is turned over. The next step is to dig away the surrounding soil until the specimen is left standing on a pedestal of earth. In the case of very large bones it is advisable to make holes at intervals through the pedestal at a level of about 1.5 cm below the bone (Fig. 4c). Dry bandages are passed through these and tied off on top of the plaster case (Fig. 4d). The pedestal is now cut away at a level below the bone and the specimen carefully rolled over. When it has been turned the temporary bandages and excess matrix are removed (Fig. 4e). Should this side of the bone require consolidation, this must be completed and the solution allowed to dry. Wet paper followed by plaster bandages and splints, where required, are applied. The ends of these bandages should be carried over to enclose those already applied by about half the circumference of the package (Fig. 4f, Fig. 5a–b, Fig. 6). The identity of the contents of each plastered package should be marked on the outside and top and bottom indicated. Smaller bones not requiring support can be wrapped in newspaper and a label attached. The removal of these should be carried out in a logical order, i.e. vertebrae in groups (cervical, dorsal, lumber, sacral and caudal) and packed in association in boxes. Ribs will probably have to be collected in fragments, and all fragments of the same rib

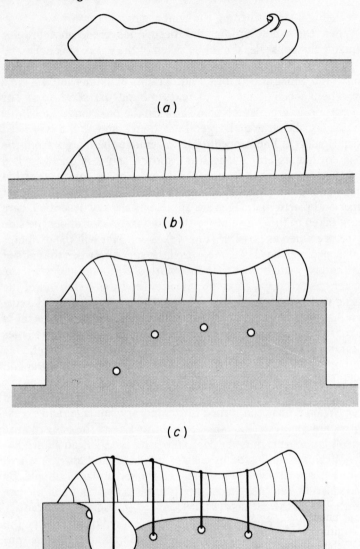

Large vertebrate specimens 49

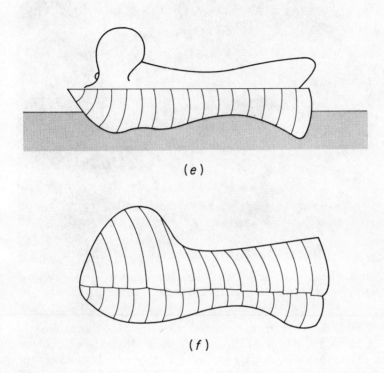

Fig. 4. Stages in excavating and plaster bandaging a bone in the field: (a) Bone uncovered to about half depth; (b) Bandages applied to this side. (Longitudinal bandages not shown; (c) Bone covered to just over half its diameter in plaster bandages, resting on a pillar of matrix formed by excavating the surrounding soil, note holes for temporary supporting bandages used in next stage; (d) Excavation continued to establish lower level of bone and temporary supporting bandages applied; (e) Half covered bone removed from pillar of matrix. Temporary bandages have been taken off; (f) Complete plaster bandaged package containing bone.

50 *Collecting and Treatment in the Field*

Plate 4. Fletton Plesiosaur. Detail of excavation showing numbering of bones. (By courtesy of Imitor Ltd. Copyright Imitor Ltd.)

should be kept together; right and left sides packed separately.

Carrick and Adams (1969) introduced an alternative technique to replace plaster bandages when lifting fossils. Instead of wet paper the specimen, when approximately half its depth has been exposed, is covered with aluminium foil of the type used in cooking. It is moulded to the specimen by hand, care being taken not to make holes in it. The foil is extended onto the ground some 5–8 cm from the edges of the specimen. A dam of cardboard or any other cheap flexible material is built round the outer edge of the foil, with walls extending about 15 cm above the highest point of the bone. The two liquid parts of an expanding polyurethane foam formula are mixed together and poured into the mould formed by the dam. The two liquids react, and rapidly produce a solid honeycombed plastic which is rigid, light and strong, and has several times the volume of the material in the liquid phase.

The soil around the foam package is dug away until it stands on a pedestal and the specimen is turned over. The newly-exposed

Large vertebrate specimens

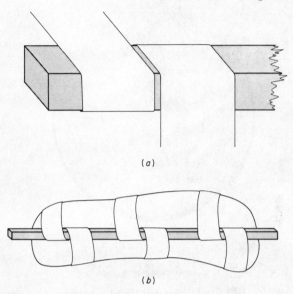

Fig. 5. Method of applying splints to the outside of a plaster bandaged package. (a) Method of attaching bandage to splint; (b) Completed assembly.

side of the bone is partially cleaned and strengthened when necessary. After the consolidating solution has dried, aluminium foil is once more applied, and a dam built onto the edges of the block of foam and held in place by sticky paper, pins or any other convenient means. More foam is mixed and poured into this.

This is an excellent method and the authors are to be congratulated on having made the only major advance in this type of collecting technique in many years. The sole disadvantage is the cost of the materials, but some of this can be written off against the saving in man-hours. Mr R. Croucher of the British Museum (Natural History) has used a foam and plaster technique as a hybrid method. In stage 1 of the excavation, he used the foam, and when the specimen had been turned, he used plaster bandages in the usual way on the underside. This halves the amount of foam used, and has one advantage over the all-foam method, in that the specimen will dry out slowly through the plastered side and this is in most cases desirable. Polyurethane foam is not only a great time

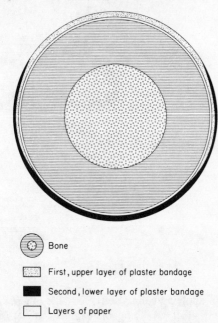

Bone
First, upper layer of plaster bandage
Second, lower layer of plaster bandage
Layers of paper

Fig. 6. Diagrammatic section through a plaster packaged bone.

saver in the field, but is much easier to remove from specimens in the laboratory. A strong-bladed knife is the only tool required.

A word of warning to new users of these materials is necessary. At least one of the ingredients may be poisonous and the vapour from it can be dangerous if it is used in an enclosed space, especially if the operator is smoking. Some can cause skin irritation. With some brands it may be necessary to neutralise the containers before they are thrown away. There is no danger if the makers of the polyurethane foam are asked for a technical leaflet and their instructions are rigorously carried out. If no hazards are mentioned in the leaflet, it would be worthwhile to make special enquiries on this point.

The pelvic and pectoral girdles of plesiosaurs and pliosaurs are made up of wide flat bones and these are very often found in a crushed and comminuted condition. Normal consolidation methods are not enough to strengthen them in the field. A technique using an acetone soluble resin adhesive like Butver B76 and a light woven glass strand mat has been developed. The surface of the bone is cleaned as much as is possible and treated with a solution of Butvar B98. The object of this is to strengthen the bone. Strips of glass strand mat are cut into convenient lengths and about 8 cm wide. Starting at any edge of the bone, after the consolidant has dried, an area just wider than the glass mat strips is coated with a solution of Butvar B76 in acetone. The solution should be viscous but mobile enough to impregnate the glass mat. A strip of glass mat is placed on top of this and liberally coated with the same solution. A stiff brush loaded with the same plastic is used to stipple the cloth tightly down to the bone surface. A small border should be carried over on to all edges of the bone. The process is repeated, overlapping slightly the strips of glass mat, until the specimen is covered. An open-woven cloth, such as muslin, may be used in place of glass fibre. The covering is allowed to dry hard. When the bone is very large the glass coated side is now covered with wet paper and plaster bandages. It is then turned over in the normal way and the underside covered with wet paper followed by plaster bandages. This side should be labelled *Bottom, remove first*. The reason for this is that when the specimen arrives at the laboratory, all packing including the glass mat has to be removed and it is easier to start from the underside. The laboratory procedure is described as a special case in Chapter 4.

Natural Moulds

In some localities the fossils have been weathered or dissolved out of the rock and all that remains are natural moulds. If these are very big they have to be collected, taken back to the laboratory and casts made from them there, but when they are small enough casts may be made in the field.

A very useful substance for this purpose is 'Vinagel 116', made by Vinyl Products Ltd. It is a putty-like material based on polyvinyl chloride. Where the natural moulds have no great undercuts this can be used to take impressions from them. The method is first to dust the inside of the mould with french chalk, a piece of 'Vinagel 116' of suitable size is then moulded in the hands. Since the plastic is thixotropic this action softens it. The 'Vinagel

116' is then pressed against a smooth surface, such as a glass plate, which has previously been dusted with french chalk. This is done to provide an unmarked surface. The smooth material is now pressed into the mould and allowed to remain for a few minutes. The plastic is then carefully withdrawn, care being taken not to distort it and the cast so formed is packed into a box in a manner which will ensure that the still soft impression will not be damaged. On return to the laboratory it is 'cooked' at 150°C until it becomes resilient and no impression can be made in it with a matchstick. At least one palaeontologist has successfully 'cooked' Vinagel in the field by heating it in a tin over a fire. This of course, has to be done with care to avoid either overcooking and burning the material, or undercooking it.

Rubber latex can be used in the field in dry weather. The latex is painted into the mould and allowed to dry. Two coats are applied, the second after the first has dried. The cast can be strengthened by applying bandage soaked in latex to the back. Any number of layers can be built up as the liquid latex, on setting, will adhere strongly to a layer which is already solid. Brushes used must be washed immediately after each coat of latex has been applied. This technique is mainly applicable to specimens in the horizontal plane. Those on a vertical plane are best dealt with with a latex thickened with aerogel silica ('Santocel' 54).

The formulae are as follows:

Mixture A	Rubber latex	300g	Included 3g of finely ground Burnt Umber in each formula to produce a brown cast.
	'Santocel 54'	3g	
Mixture B	Rubber latex	300g	
	'Santocel 54'	12g	

The 'Santocel 54' is mixed a little at a time into the rubber latex. The mixing may be done by hand in a mortar using a pestle and a steel spatula, or with practice, a laboratory stirrer may be used. Mixture A is not much more viscous than ordinary latex; mixture B is a spreadable paste which will not run off vertical surfaces. Both mixtures should be made up several days before they are required to allow air bubbles formed in the making to disperse. Like all rubber mixes, both shrink on drying. Mixture A is first painted into the mould and when dry mixture B is spread on to thicken the cast.

As has been previously mentioned 'Quentglaze' can be used to consolidate impressions in sand and pervious wet soils.

Zoologists have used polyester resins to make casts of recent molluscan burrows in the field. Epoxy resins would be better as they are less affected by water. This technique could have applications in palaeontological field work.

Nodules

Nodules occur in many kinds of rock and often contain fossils. Because they are harder than the surrounding rocks, they can weather out and are sometimes to be found at the bottom of banks and cliffs or on beaches. If there is no evidence of a fossil showing on the outside, it is best to split the nodule open in the field, but where there are signs of bone, however, it is better not to do this, but to take the whole specimen back to the laboratory. When a split nodule contains a fossil, both halves must be consolidated if any part of it is loose and when dry, packed carefully to avoid damage to the exposed specimen in transport. It helps if each half is numbered with the same number, and, should the nodule have broken into more than two pieces, the numbers should be suffixed by a letter.

Many fine vertebrate fossils have been found contained in nodules from the Lias of Dorset and other places. Some very small ones, found in a Miocene beach deposit in America, on solution in hydrochloric acid, were found to contain a wealth of insect remains most beautifully preserved in transparent silica.

Fossil Plants

The study of the remains of fossil plants requires the use of highly specialised laboratory techniques. For this reason, if it is known that fossil plants are likely to be found before going into the field, a palaeobotanist should be consulted as to what may or may not be done to them. His answer is likely to be, 'Nothing but careful labelling and packing.' The use of consolidating solutions on leaves and other parts is generally not permitted. The collection of columns of matrix or borings for pollen analysis are jobs best left to palaeobotanists or their assistants who have been specially trained for this work. Some idea of the complexity of the study of fossil pollen and spores is indicated by the fact that Part III of the *Handbook of Palaeontological Techniques* (1965) is devoted entirely to palynology and is 234 pages long.

Matters of General Importance

The collection of specimens is only the first step in their preparation for study or display and the collector should bear in mind that what he does in the field will very often have to be reversed in the laboratory. This is always true of plaster bandaging or any like form of field packing. If it is remembered that what goes on must also come off, little trouble will arise. Many collectors overestimate the amount of plaster bandage which is required to ensure the safe transport of the specimen. It is true that a little too much is better than not enough, but let it be remembered that the operative word is 'little'. Plaster bandages are very easy to put on, but they are not so easy to remove without damaging the specimen. A mistake to be avoided is the introduction of a mass of plaster and hessian into undercut hollows in the bone or between bones. This commonly happens when large mammalian skulls are collected. The space between the skull and the zygomatic arch tends to arrive at the laboratory filled with an exceedingly tough mixture of plaster and hessian. This is never necessary. A way to avoid introducing plugs of plaster and hessian into unwanted and unnecessary places is to pack the hollows, before the plaster bandages are applied, with wet paper. For this purpose newspaper is better than anything. Although a mammal skull has been cited as an example, these remarks apply to any type of fossil which has to be encased in plaster bandages in the field. They apply with even more emphasis if old sacking of a strong and tightly-woven type is used in place of plasterer's scrim or hessian.

Safety in the field is largely a matter of common sense, but enthusiasm can often result in the obvious being overlooked. All hammers and picks should be checked to see that their heads are securely attached to the shafts, which themselves should be inspected for signs of weakness. Some specimens are so heavy that lifting tackle is required to move them; attempts to lift such fossils without the proper equipment can only end in disaster.

Operations such as tunnelling into sheer faces should never be attempted. When the specimen disappears into such a face, and the overburden is so great that it cannot be dug away, there is nothing you can do about it without professional help from people who are trained to do this sort of thing with safety.

The laws of trespass apply as much to the fossil collector as they do to anybody else and a great deal of unpleasantness can be avoided if the landowner is approached before collecting is started on his property. A word of thanks, when the expedition ends, is never out of place, as it will not only make things easier if you wish to return, but will also increase the recipient's goodwill towards palaeontologists in general.

Field Labels

The importance of correctly labelling specimens as they are collected has already been stressed, and because preservation of labels is vital it is perhaps advisable that a little more should be said on this subject. Equal in importance to the absolute accuracy of the information contained on a label, is the necessity to ensure that it cannot be destroyed or removed from the specimen easily. Palaeontological laboratories are busy places involved in all kinds of projects, ranging from assisting scientists in their researches to mounting large skeletons for exhibition and, therefore, must have a work plan. Although the staff may be eager to start preparing a specimen as soon as it arrives from the field, their work load may make this impossible and the fossil may remain in storage for years before it receives attention. Labels must therefore be capable of surviving for a very long time and should, ideally, be written in waterproof ink. When this is not available it is better to use a medium hard pencil rather than an ink which will smudge and become illegible if wetted.

Linen labels are made for various purposes and are less easily damaged than paper but even these, if there is only one attached to the exterior of a package, can be torn off and lost. It is safer to have two labels for each specimen, one inside the package and the other outside. Collecting bags are often sold with a label attached to the neck, but it is still advisable to put a copy inside the bag to ensure against loss. Specimens are often packed in small cardboard boxes and a common mistake is to write the applicable label on the box lid only. Lids can become separated from boxes even if they have been secured with adhesive tape, it is, therefore, much safer to write the labels on the lid and the box and doubly so if once more a copy is packed inside.

It has been suggested that when a fossil is enclosed in plaster bandages the field information should be enscribed in the plaster, this would appear to be absolutely safe but it is not. Plaster encased bones may be moved several times, while in storage before they are unpacked, and the plaster can become eroded and the label obliterated, so again a label should be placed inside this type of packing. If the contents are blocks of firm rock, information can be painted on these before the plaster is applied. In any case, labels written in pencil on pieces of paper which will not disintegrate when wetted can be placed on the specimen before the layer of wet paper is applied. To be absolutely safe there should be more than one copy in each package.

A practice to be avoided at all costs is the writing of labels on packing paper and leaving it at that. This is most unfair to the person who may be called upon to unpack the collection. He may be a junior who is not aware that this dangerous form of recording information is sometimes used by collectors, and throw the paper with the data away, thus rendering the collection scientifically useless. It may be argued that such a catastrophe would be the fault of his superiors, and it is quite true, but nobody is infallible and it is quite easy for a person of experience to assume that because he would automatically check all paper wrappings everybody else will do the same.

References

Camp, C. L. and Hanna, G. D. (1937) *Methods in palaeontology*, Berkeley: University of California Press.

Carrick, J. N. and Adams, S. J. (1969) 'Field extraction and laboratory preparation of fossil bones and teeth using expanded polyurethane', *Proc. Geol. Assoc.*, **80,** 81–9.

Colbert, E. H. (1965) 'Old bones and what to do about them', *Curator*, VIII, 302–18.

Kummel, B. and Raup, D. (eds.) (1965) *Handbook of palaeontological techniques*, San Francisco: W. H. Freeman & Co.

4
Mechanical Development

The removal of the matrix from a fossil is known as 'development'. This term has been in use for so long that, although it is not a good definition of the process, it will be retained here for want of any better. Matrix may be removed either by mechanical or chemical means. No matter which method is employed there are certain things that must be done before work is commenced.

Specimens may arrive at the laboratory from a number of different sources. They may have been in collections for many years and require further development because, in their present state, they do not show enough detail to satisfy a scientist researching into the group of fossils to which they belong. Others may have been sent in as enquiries by a member of the general public or by a university department and some will come directly from the field enclosed in plaster bandages or blocks of plastic foam. Irrespective of their source the first thing which must be done is to ensure that they are accompanied by a label, if one cannot be found, immediate enquiries must be made concerning the locality from which the specimen was collected, who collected it and when, whether it has been identified and registered and whether the whole of the specimen is there. These enquiries will take time, but it is not wasted, and no preparator should ever allow himself to be dissuaded from making them. When the information has been collected it must be recorded at once and written on labels which will remain with the specimen throughout its development and be passed on to the person responsible for the safe custody of the specimen. When there is no trouble of this sort, the original labels with the specimen must be removed, preserved and copies, which are kept with the specimen at all times, made.

When dealing with specimens which are encased in plaster bandage or plastic foam it is obvious that the packing must be removed before any preparation can be carried out. It is rarely possible to preserve the areas of plaster into which a field number has been inscribed, so great care should be taken in copying it.

When removing plaster bandages it is easier to take off first those which were put on last. If the collector has marked the packet 'top' and 'bottom' the question as to which were the last applied is answered at once. It must be on the side marked 'bottom'. Sometimes a collector does not mark his packets in this way, but provided that he has not overdone the plastering, it is fairly easy to see which bandages were applied last. They will be those whose ends are visible on the surface. The removal of the bandages is made easier if the plaster is softened by wetting. Warm water, with a little detergent, may be applied with a sponge, or it may be sprayed on.

There are several makes of household cleaning fluids which are sold in plastic bottles fitted with very efficient sprays and such bottles, when empty, are worth collecting, for they can be used for damping plaster and many other laboratory operations, and are less messy in use than sponges.

The tools required to remove plaster bandages are a strong, sharp knife and a hacksaw blade fitted into a pad saw handle. The knife is best kept sharp by a coarse hone of the type made for sharpening gardening tools; a razor edge is useless as it will blunt at the first stroke. The method is to start very carefully in order to ascertain what sort of condition the specimen is in. When weak, it may have to be uncovered a little at a time and strengthened as the job proceeds; if it is strong and the man in the field has done his job properly, large areas can be uncovered at a time.

As was mentioned in the chapter on collecting, some field workers grossly overdo the plastering thus making a much larger tool kit necessary for the removal of the packing. A thick layer of plaster may have been smoothed all over the package, making it impossible to see where the bandages finish. The safest way to deal with this situation is to wet the plaster surface thoroughly and cut it away with gouges until a layer of bandage is revealed. When enough bandage has been uncovered, cut through it with the knife and peel it off. The cycle is repeated until eventually the paper layer is reached. Sometimes the task is so difficult that powerful pincers and shears are required. The most important thing is to remain patient and not resort to using a hammer and chisel.

The reason why it is advisable to uncover the bottom first, i.e. the part which was removed from the ground last, is that this is most likely to have received the least attention in the field and will have more matrix on it than the side which was first uncovered. This is a matter of little or no importance if the specimen is a bone or an association of bones in a solid block, except that it is sometimes safer to uncover the barren side while the specimen-bearing surface is still protected.

In the case of long bones or skulls taken from a soft matrix, it matters a great deal that the weakest side should be dealt with first. The specimen should be cleaned and consolidated as much as possible as the bandages are removed. When the exposed side has been made as strong as possible the remainder of the bandages should be removed by stripping them off from the sides and towards the surface of the bench, consolidating newly exposed areas as necessary. It is convenient if all specimens requiring this treatment are placed on boards on top of the bench rather than on the bench surface, for they can then be moved to the easiest working position without any force being applied to the package itself. When all the bandages have been stripped down as far as possible, all consolidating solutions have dried and if the specimen is now strong enough, it can be carefully rolled out of the remaining bandages. Very often there will be major fractures through it, and in this case it is better to remove the pieces separately. They should be taken out starting from one end and laid out in their correct relationship, one to another, on a board. Repairs and any further development are then carried out. The special case of the flat and badly damaged limb girdle bones of marine reptiles is dealt with at the end of this chapter.

Traditional Methods of Mechanical Development

The earliest professional preparators were, almost without exception, men who had been trained as stonemasons; for this reason most of the development was done with hammers and chisels and it was not until the first quarter of this century that even the dental hammer was used. The great skill of those men is evident from the hundreds of beautifully prepared specimens which are to be found in museums throughout the world. Even today it is necessary that all preparators should master the art of using hammers and chisels, for there are still many specimens which cannot, in spite of the great advances made in the last thirty years, be developed in any other way.

Hammers and Chisels

The best type of hammers for this work are the two-faced kind varying in weight from about 150g to 1kg. Those which are made for bricklaying or allied trades are preferable to all others. Sculptor's hammers can be used, but these tend to have sharp corners and can be a source of pain if the preparator has not reached a level of skill where he never misses the chisel and hits his hand.

Some of the chisels needed can be bought ready made, but most of these are too broad for delicate work. This means that the preparator must acquire yet another skill; that of being able to forge and temper his own tools. This is not too difficult, where straightforward articles like chisels are involved, but it is worth extending one's ability beyond this, for very often the preparator's work is made easier if he is capable of making tools designed for a special purpose. Instructions for forging and tempering chisels are given in Appendix I.

Development by Hammer and Chisels

When a specimen is to be developed by hammer and chisel it must be held in some way so that its movement will be restricted under the constant rain of blows to which it will be subjected. The chisel has to be struck with considerable force when the matrix is very hard and the shock waves created can break the specimen if it is not supported on the underside by some resilient material. Very large and heavy specimens need only to be placed on a sack partially filled with a wood wool or alternatively a thick sheet of foam rubber. If the specimen is flat its own weight will hold it in position. Irregular shaped fossils may be held by placing heavy blocks of iron, about 30 cm in length against their sides: the iron must, of course, be padded. Strongly made wooden boxes, 30 cm × 10 cm × 10 cm, filled with sand and closed with a screwed down lid, are an alternative to iron blocks, and have the advantage that they can be made very cheaply. Smaller specimens can be placed on top of a tightly woven cloth bag which has been partially filled with sand. The fossil will bed down into this and is further steadied by the preparator's wrist or forearm.

Nowadays specimens of up to 10 cm in length are not developed by hammer and chisel; one of the mechanical tools to be mentioned later (see pp. 71–80), is used. Before these tools were available the specimens were, if surrounded by enough matrix, held between the parallel, wood-lined jaws of a carpenter's vice. When there was not enough matrix present to make this safe, the fossil was covered on one side with wet paper and embedded to this level in a plaster block and this was stuck to a heavy slab of slate.

After the method of supporting the specimen has been decided, it is examined to establish the best side on which to commence working. In the case of vertebrates this is always the side with the most exposed bone; in that of invertebrates it is the surface which is least covered with matrix. All weak areas of visible specimen are strengthened by consolidation and the consolidant allowed to dry. The chosen side is now placed uppermost. Taking the case of a block containing vertebrate fossils, an area of exposed bone is selected and holding the chisel in the left hand at an angle of about 60°, the blade or point is placed on an edge of matrix about 3 mm or less away from the bone surface. The left wrist should rest on the block and be tensed in a manner which tends to pull the chisel back from the bone. The butt of the chisel is now struck with a hammer and a small flake of rock should spring off the bone. The chisel is moved sideways and the process repeated until the width of the specimen has been traversed. It is then moved back a little and the action repeated, back and forth, until the specimen is completely uncovered on the top and sides.

The chisel is driven towards the bone because, as a rule, there is a plane of weakness at the interface between the matrix and the specimen and the rock will split off cleanly there. If the chisel is driven away from the bone, i.e. an attempt is made to lift off the matrix, almost certainly a piece of the bone surface will be pulled away with the fragment of rock. Every effort should be made to avoid marking the surface of the fossil with the chisel. The pulling-back action of the wrist, mentioned above, tends to stop the blade touching the bone. The hammer blows should be as light as possible, exerting only enough force to produce the splitting action required.

Should a piece of bone come away attached to a flake of rock the scar and the detached fragment should be consolidated with a quick drying solution, such as Butvar B98 in isopropyl alcohol and when this has dried, the piece should be stuck back accurately with polyvinyl acetate emulsion. After the emulsion has hardened the matrix may be removed from the bone fragment by the careful use of a very narrow bladed chisel and a light hammer. The emulsion is used because, when dry, it remains resilient and the joint will not break while the area above it is being developed.

When during development, the newly exposed bone shows signs of weakness it is consolidated and no more work should be done in this area until it is dry; it can be lightly covered and operations transferred to another part of the specimen. The point has already been made that the removal of very hard matrix requires that the hammer be used with considerable force. The effect of the shock waves produced on bones in all parts of the block should be constantly observed, for it is quite possible to jar a specimen to pieces at a point some distance from the one being developed if this is not borne in mind.

Thin or narrow structures, such as neural spines in the vertical plane are best developed by working round them from top to bottom. This ensures that the force is always applied at or near the base of the object and so is less likely to snap it off than if it were exerted at a higher point when a leverage component would be introduced. When, for one reason or another, one side of such a structure has to be cleared first, it must be supported before any attempt is made to expose the other. In the past this would have been done with plaster of Paris but now, provided that the specimen is not so large as to make this uneconomical, polyethylene glycol 4000 is preferred as it is much easier both to apply and remove.

The common practice in the mechanical development of associated bones is to develop one side only because, in a hard matrix, it is virtually impossible to separate them without damage. Since one side of some bones is the mirror image of the other, these specimens are developed to show one complete side of each bone. A string of articulated vertebrae, therefore, shows one side of the neural spine and arch, one pre- and one postzygapophysis, one transverse process and half of the centrum of each. How much of any other bones associated with the vertebrae can be revealed is entirely dependent on what happened to the skeleton before it became firmly embedded in the rock. This is the great limitation of mechanical development, for very rarely is an associated skeleton embedded in such a way that mechanical development can reveal every anatomical detail. This can also apply to isolated joints and lead to misidentification. For something like a century the type specimen of *Scelidosaurus harrisoni* was an isolated knee joint which had been prepared by hammer and chisel; Newman (1968) become suspicious of this and after it had been prepared chemically and the individual bones disarticulated, it proved to have nothing to do with the herbivorous ornithischian *Scelidosaurus*, but was the knee joint of a carnivorous saurischian dinosaur.

The complete removal of matrix from the outer surfaces of isolated skulls of large to medium size is easily achieved with hammers and chisels. If the lower jaw is present it is stronger while it is still supported by matrix as are the upper teeth if the lower jaw is missing. For these reasons it is best to clear the top and sides of the skull first. The support to the underside should be augmented by placing it onto a pad of thick foam rubber. This is normally placed on a board so making it easier to turn the job round than if the rubber were in direct contact with the surface of the bench. When the top and all sides of the skull are free, it can be turned over and the underside cleared. Where lower jaws are present and are articulated, it is not possible to remove them; this makes the development of the insides of the teeth and palate very difficult. The use of mirrors or a sheet of mirrorised 'Melinex' helps to direct the light to the right places and makes it possible to see underneath projections. Dental picks are useful on medium-sized skulls, but for larger ones special tools very often have to be made.

This method of approach is valid for normally-shaped skulls which have not been artificially flattened by earth pressures. Skulls which have little depth in comparison with their length and breadth have to be supported, after one side has been cleared, before work begins on the other. The long-snouted crocodiles such as *Steniosaurus* and some other aquatic reptiles, have skulls which are of this form. In these cases, whenever possible, the top of the skull is developed first and is then embedded in some rigid substance. The traditional procedure is to cover the exposed bone with wet tissue paper and then with plaster of Paris. The natural cavities, orbits, nares, foramen magnum, etc., are packed out with wet newspaper to prevent the formation of blocks of plaster in them. The skull is placed on a sheet of plate glass which has been oiled and plasticine packed round it, so that the plaster cannot run under the specimen. A dam of plasticine is built round the specimen at about 2.5 cm from the edges and the liquid plaster is poured to a depth sufficient to provide a block strong enough to withstand the hammering; the base of the plaster block is flattened and allowed to dry out. Plaster of Paris is fairly brittle when dry and to give added strength, a piece of wood about 2.5 cm thick is stuck to the base.

Development of the rock-covered side now begins. The preparator should at first work in one area to establish the level of the bone and when it is found, work outwards from here. After the underside has been completely cleared, the plaster block has to be removed. This is best done by wetting the plaster thoroughly on all sides to soften it and cutting it away with gouges until a point is reached where it no longer grips the specimen. A piece of foam rubber is placed over the fossil and, on top of this, a piece of wood which is slightly longer and wider than the foam pad.

Lengths of string or sticky paper tape are passed under the plaster block and fastened on top of the wood; then the whole is turned over. The string or tape is cut and an attempt is made to lift the plaster off the skull. No force should be used; if the separation is not easy the wooden backing of the plaster block should be removed by sawing carefully through the plaster. The specimen can then be freed by cutting away the remaining plaster with gouges.

Polyurethane foam can be used in place of plaster but in this case, the specimen should be given a coat of polyvinyl alcohol solution. When this is dry, aluminium foil is used in place of wet paper to keep the foam away from the bone surface. Polyethylene glycol wax 4000 is ideal for support purposes, but since the wax is soluble in water the dam must not be made of clay. The wax may be applied directly to the bone surface. It is inflammable and must be melted in a double saucepan on a low heat electric hot plate and should not be raised to a temperature much above its melting point, which is about 50°C. There is a tendency for voids to be formed in this wax when it cools if it is poured at too high a temperature, so when the block has solidified it should be probed with fine needles to ascertain that no such cavities are present. When they are, the area is excavated to give access to them and they are filled with more wax. The wax block should be backed with a piece of hardboard or wood. This can be stuck to the main block by first warming it, covering the rough side with molten wax and then applying it quickly to the main block. To remove the wax, when the job is complete, it is cut away at several points close to the specimen and pins are fixed into the board. The specimen plus the remaining wax is now placed in a large flat dish, the volume of which is greater than that of the wax, and one end raised on a block of wood so that the specimen is inclined at about 15° to the horizontal. The whole is placed in a heat-controlled electric oven which is set to a temperature slightly above the melting point of the wax. The wax will melt and the majority of it will flow down into the tray. The pins will stop the specimen sliding off the board. The specimen and board are removed from the oven and the fossil can then be transferred from the waxed board to a clean one. Most of the wax which remains on the specimen can be removed by careful work with a steel point and, what is left after this, by washing in warm water.

The use of polyethylene glycol 4000 is best confined to smaller specimens because of its relatively high cost. Although it can be recovered and used again, as shown above, large specimens require the use of large ovens which must be of a heat-controllable type with no flame or red-hot elements inside the oven chamber, and such ovens are expensive. When working on small fossils it is convenient to have a few cubic centimetres of molten polyethylene glycol 4000 readily available. Apparatus designed for this purpose is described in Appendix I.

Tissues other than Bone

Throughout the description of development by hammer and chisel, the word 'bone' has been frequently used, because the overwhelming majority of vertebrate fossils in matrix with which the preparator will have to deal will be represented by bones alone. Very rarely vertebrate fossils are found in which the outer parts of the body are preserved. The skin impression of reptiles, and sometimes the skin itself, is preserved and should be looked for. Fossil fish commonly have their scales intact and because the junctions between them are filled with matrix, these require more than usual care in development. Pointed, rather than chisel-ended tools should be used and only a little matrix removed at each hammer stroke.

All the methods of mechanical development described apply to invertebrates as well as vertebrates.

The Development of Small Fossils

The development of small vertebrate remains, especially skulls, and highly ornamented invertebrates should always be done under a binocular microscope. The microscope chosen should have an optical system which gives a good working distance below the objective, and preferably be fitted with a zoom lens. It should be mounted on a long arm which is attached to an upright seated into a heavy base. A variety of eye pieces is needed, $\times 5$, $\times 10$, and $\times 20$ being a good selection and divide-by-0.5 lens for attachment to the objective extends the range. The middle-priced microscopes are often better for this purpose than the very expensive models with refined optical systems. In all cases a good source of light is required; a microscope lamp with a focusing device, powered by a low voltage variable transformer and capable of being used freestanding or attached to the body of the microscope, is ideal for most of this work. It is an advantage if the lamp is provided with some means of carrying coloured filters.

To develop a small fossil under a microscope, it should be held in one hand. When it is too small or too delicate to be held in this way, or requires support to withstand the forces used in mechanical development, it is partially embedded in a block of polyethylene glycol 4000 attached to a piece of hardboard. This is done in the following manner. A small piece of hardboard is cut so that it is large enough to accommodate the specimen with a generous margin of wax round it, and still leave enough space for a dam to be built on it to enclose this area. The dam is built of plasticine and into it a small amount of molten polythylene glycol 4000 is poured. The fossil is placed in this, with the side to be developed upwards and after the wax has set, more is poured in up to the level required. Should the polyethylene glycol 4000 overflow onto the top of the specimen it is a matter of little importance, as it can easily be removed with a needle after it has solidified. Before work starts the block should be examined for cavities and any that are found filled.

The simplest tools for use on small specimens are needles or small steel points held in different sizes of pin chuck. Ordinary sewing needles, which can be bought cheaply in many sizes, are best for light work; heavier work can be done with 15 s.w.g. headless steel pins which can be obtained in the United Kingdom, from the Armstrong Cork Co. in a variety of lengths. The most suitable length of pin is about 3 cm ($1\frac{1}{2}''$) since this places the point further away from the body of the chuck, and so the view of the working area is not obscured. The splitting force required to separate the matrix from the specimen is applied by finger pressure of the hand holding the tool; the needle point is driven towards the surface of the specimen, so splitting, not dragging off the rock. Dust and loose pieces of rock can be conveniently removed with a rubber blow-bulb of the type obtainable from the manufacturers of chemical apparatus. Needles may be used straight, but some preparators prefer to bend them slightly at about 1.5 cm from the point. The advantage of this is, that when the needle is used with the curved edge towards the specimen and it slips, the point will travel upwards and is therefore less likely to dig into the specimen and damage it. During preparation the points of the needles become blunt and can be re-sharpened on a two sided axe-stone. This type of whetstone is round and has a coarse abrasive on one side and a fine one on the other. The stone may also be used to make a chisel point on a needle or steel pin.

As in hammer and chisel development, any cracked or weak bone must be consolidated and allowed to dry before any more work is done near it. The consolidating solution should be applied with a small brush under the microscope, and any excess removed with solvent. This is sometimes a very delicate operation on small specimens, especially if they are highly ornamented or small teeth are present. The secret is to keep the amount of solvent down to an absolute minimum and to avoid any general flooding of the area.

Some matrices are easier to remove if they are wetted and, generally, this tends to accentuate any colour difference between them and the fossil. Obviously the surface of the fossil must be dried before any consolidating solution is applied. When the specimen is stable enough to stand it, the drying can be speeded up by blowing air across it with a blow-ball. The wetting technique is particularly useful when developing out the hinge teeth of lamellibranchs. When these fossils are preserved in matrix which is hard and differs little in colour from that of the shell in the dry state, wetting them will often produce a marked contrast. Liquids other than water can be tried; toluene or xylene will often succeed where water fails. Contrasts have also been achieved by using different coloured light filters.

Commonly the matrix containing the lamellibranch has pieces of shell scattered through it which are not part of the main specimen, and these can cause confusion if they are in the region of the hinge teeth. The only safe policy to adopt is to assume that they are part of the hinge until it is proved beyond doubt that they are not, and then they can be cut away. The methods mentioned for obtaining contrast are very helpful in these cases. When removing matrix from hinge teeth of shells, in which the two differ very little in colour and hardness, there is a danger that, if the preparator concentrates too much on what he thinks the hinge should be like, he will subconsciously carve a very convincing artifact. In such cases it is best to start work from an area of undoubted shell and work away from this, checking that every newly-exposed area is shell and never making a guess as to what will happen next.

Nobody is infallible, and small pieces may be knocked off a fossil in the course of development. No matter how small these are, they can always be put back if they can be found. The chances of their being retrieved are increased if a large pad of foam rubber covered with a white cloth is placed on the bench below the microscope, and is kept as free as possible from loose chips of matrix. When found, a fragment can be picked up on the moistened point of a fine paint brush, and placed back on the fossil as near to its correct position as possible. If a fairly slow-drying adhesive is applied to the broken surface on the specimen, the fragment can be manoeuvred back into position with the point of a needle. This may take some considerable time during which the adhesive may dry out. It can be softened again by the careful application of a suitable solvent. Sometimes loose pieces can be floated back into position in a solvent onto a patch of adhesive which has nearly dried out.

Modern Tools for Mechanical Development

Some matrices are too hard to be removed by finger pressure applied to a needle and some fossils are too small to be developed by hammer and chisel. There are a number of tools currently available which help to overcome these difficulties.

The Dental Machine

This is the same apparatus as used by dentists to drill and fill teeth. The small portable types are not expensive, and a variety of stands are available for supporting the motor. The handpieces are of two types, rotary and persussion. Rotary handpieces may be of light duty or heavy duty type. The heavy duty type is made for dental mechanics and is harder wearing than the light duty form, but there is the disadvantage that it has a heavy chuck and is rather too clumsy for delicate work. The light duty handpiece is made for dentists but is not designed to withstand long running periods, and will overheat unless it is rested at intervals.

Both handpieces can be used with a wide variety of diamond cutting wheels and different shaped grinding burrs. Synthetic abrasive burrs are much cheaper and sometimes more effective. Circular diamond burrs which have the flat face as well as the edge charged with diamond are most generally useful. The synthetic abrasive type in the form of a truncated cone with the base forward are a suitable substitute for these. A variety of brushes made of steel and brass wire, as well as bristle, can be fitted into the handpieces. All dental suppliers issue illustrated catalogues of these tools and the preparator can make his own choice from them. Although the price of cutting tools does not vary a great deal, the price of handpieces does, and for palaeontological work it is better to buy the cheaper ones, this is because the tools are not being used for the purpose for which they were made and for our work the cheaper types last as long as the more expensive.

Dental burrs and cutting wheels are used mainly to grind down the matrix on a fossil until it is thin enough to be removed by percussion. The method is to find an exposed area of the specimen and grind the matrix towards this until the area about it is thin enough to be split off. When, by this means the thinned area of matrix has been removed, grinding is repeated to reduce the thickness of rock surrounding the newly-exposed part of the fossil. Some workers advocate grinding the rock right down to the surface of the specimen but this is a hazardous proceeding, as it is extremely easy to take it too far and grind the surface of the fossil as well. The grinding followed by percussion technique is one of two possible methods of developing those fossils which are enclosed in haematite. This method, combined with the use of polyethylene glycol 4000 to support some very thin edges of bone was used to develop the skull of the type specimen of the ornithischian *Heterodontosaurus* in the British Museum (Natural History).

Air turbine dental drills, although they are very many times more expensive than those driven by motor, work at high speed, but will stall immediately if too much pressure is applied to the burr. This makes them safer to work with, as the stationary tool cannot skid off the matrix and damage the specimen.

The percussion handpiece is variously known as the Power's mallet, Stensiö hammer and the dental mallet. It can still be bought under the last name through dental suppliers. All but the Stensiö hammer are designed to take a variety of dental points. The Stensiö hammer is made especially for palaeontology and is fitted with a small chuck which will hold steel pins of the old gramophone (phonograph) needle type; it is rather stronger than the other varieties, but may not be easy to obtain. A small chuck can be made to fit an ordinary dental mallet and this is an advantage. The force of the blow of all types is adjustable and the rate of delivery is controlled by the speed of the motor, which is varied by a foot control. These are fairly delicate instruments and can be easily damaged. There is not much which they can do which cannot be done with the more robust 'Speed Engraver'.

The Burgess Speed Engraver Model 72 and Model 172

For approximately twenty five years the Burgess Industrial Vibrotool has been one of the most useful aids in the mechanical development of fossils of all types, but now it is no longer in production and has been replaced by two models of Speed Engravers made by the same firm.

Whereas there was no doubt at all that the Industrial Vibrotool was superior in every way to the pattern sold for handicrafts, when used to develop fossils, there is so much difference between the Hobbyist Kit, of which the engraver 72 is a part, and the Model 172 which is sold with four identical extra hard points for use in industry, that it is difficult to give advice as to which is the better buy.

The Model 72 Hobbyist Kit is built around a two-speed engraver, having a high speed setting of 100 strokes per second and a low speed of 50 strokes per second, and contains, as well, an assortment of attachments which are of no use for developing fossils, but could be very useful for making small mounts.

The Model 172 Single Speed Industrial Engraver operates at 100 strokes per second and is equipped with a special hardened point for engraving on metals and glass; and while this could be used to develop large fossils of uncomplicated structure it is much too coarse for fine work on delicate specimens. To make delicate work possible the Speed Engraver must be fitted with a chuck capable of holding a 15 s.w.g. headless steel pin. Two firms in the United Kingdom, whose addresses are given in Appendix II, used to sell Vibrotools fitted with this type of chuck, and it is reasonable to suppose that they will continue this policy with one or both models of the Speed Engraver. Should this not be the case, it is not difficult to make a suitable chuck.

The main requirements are that the jaws of the chuck should be attached to a solid tang, which has the same diameter as one of the standard points and is long enough to project for approximately 3 mm beyond the retaining collar which holds the tools in position in the Speed Engraver shaft. The jaws must be bored so that, when closed, they grip the pin tightly and its base rests on a solid

74 Mechanical Development

Plate 5. Speed Engraver Model 172. (By courtesy of Burgess Power Tools Ltd.)

shoulder in each element of the jaws, or upon the end of the solid tang to which they are attached. It is best to have chucks made by the dozen for not only is this cheaper in the long run, but also it prevents the tool from being out of action when a chuck is broken or wears out.

A cruder method of holding the pin is to take a piece of iron or mild steel rod, whose diameter is the same as that of a standard point for the engraver and whose length is such that when attached to the tool about 1 cm projects beyond the retaining collar. Into one end of this rod a hole is bored approximately 1 cm deep and of a diameter which just allows the insertion of a 15 s.w.g. pin. The pin may be secured either by drilling a hole 0.5 cm from the bored end and tapping this to take a small grub screw, or it may simply be held in place by filling the longitudinal hole with molten shellac and pushing the pin into it; when the resin sets the pin is held and may be removed by heating the iron rod in a bunsen flame and pulling it out with pliers. The steel pins used should be approximately 3 cm ($1\frac{1}{4}''$) long.

The Speed Engraver is used by placing the vibrating point lightly on the matrix, the tool being held like a pen. Only very light pressure should be applied, as the blow delivered by the engraver provides the splitting action required to separate the matrix from the fossil and too much pressure can cause the point to pierce the surface of the specimen and damage it.

These new tools have been designed to reduce weight and for easy handling. The Single Speed Engraver Model 172 is probably the more robust and since 100 strokes per second is about the speed required for the development of most fossils it should be suitable for the majority of workers. On the other hand the wide range of tools included in the Model 72 Kit, plus the choice of two speeds may be more attractive to others.

The speeds quoted are for operation from 50 Hz, AC mains. In countries other than the British Isles the mains frequency is often 60 Hz, which probably means that the speeds in these places would be 60 and 120 strokes per second; this is a point which should be checked with the makers. Both models are available for operation at 240 V, 220 V, 110 V, and 100 V AC only.

The manufacturers have a high reputation for the quality of their goods and if the new tool is anything like as good as the old Vibrotool it will give excellent service for years without requiring attention, and most breakdowns will be due to the tool being dropped. Spares for the Vibrotool were available from the makers and supposedly they will pursue the same policy with the new tools, and a stock of these saves time if a tool is damaged, but repairs should be carried out only by a qualified electrician or somebody with more than a passing knowledge of electricity.

The two models are very similar and the single speed model is illustrated in Plate 5. I have to thank Messrs Burgess Power Tools Ltd, both for the photograph and for permission to publish it. I am also indebted to them for technical information.

The S.S. White Industrial Airbrasive

First described by Stucker (1961) for use as a development tool, this apparatus works by firing a fine stream of abrasive, propelled by an inert gas under pressure, at the matrix to be removed. Straight and angled handpieces are available fitted with nozzles with different-shaped holes. The manufacturers supply the

76 Mechanical Development

Plate 6. S.S. White Airbrasive, with glove box and gas cylinder arrangement at B.M. (N.H.) Paleontology laboratory.

abrasives for use with this machine and no others should be used. There is a choice of grades of aluminium oxide powders, which cut at different speeds, as well as finely powdered dolomite and sodium bicarbonate. The rate at which abrasive issues from the nozzle is controllable and there is a quick-acting foot-operated switch for starting and stopping the stream. Air can be used as a propellant, if it is suitably filtered and dried, but most laboratories use compressed carbon dioxide and some nitrogen. Oxygen must not be used. Whichever permitted gas is chosen, the cylinder must be fitted with a reduction valve to reduce the imput pressure of the gas to eighty pounds per square inch (5.6 kg/cm^2).

It has been found that the dolomite powder is the most useful abrasive, as it is safe on most fossils. Chalk-covered echinoderms can be cleaned with this at amazing speed and all features revealed. Highly ornamented plates of fossil fish which had first been developed in acetic acid, but were still partly obscured by insoluble matrix in the ornamentation, were perfectly and very quickly cleaned with the Airbrasive, using dolomite. Before an

Airbrasive was available, it took a highly-skilled preparator many days to complete the cleaning of such specimens.

The aluminium oxide powders must be used with care as very often they will cut into a fossil more easily than they do into the matrix, and, if the nozzle is used with a sweeping movement, the fossil may be polished and such damage obscured. Owing to the high velocity of the particles leaving the nozzle they rebound at great speed from any surface they strike, and this can lead to damage to a specimen when working in hollow zones, and to the nozzle itself if this is held too close to the specimen.

It is quite obvious that an apparatus firing out finely divided powders cannot be used in the open, but work must be carried out either in a glove box or under a very efficient dust-extraction hood. The makers sell dust-extracting apparatus, but it is quite easy to build one to suit the special needs of the palaeontologist. An easily made extraction system is described in Appendix I.

A low power microscope can be used with this apparatus and either a strong desk lamp or a spotlight is needed to provide illumination. Both, of course, are mounted outside the box and the specimen is viewed through the glass lid. Perspex sheet may be used in place of glass; it will not break very easily but it will scratch.

Since more than one abrasive can be used in the Airbrasive it is advisable to have fixed to it an indicator saying with what it is charged.

Ultrasonic Apparatus

Ultrasonic apparatus used in preparation falls into three categories.

First, ultrasonic washing tanks are made by many firms in many countries and vary in size from 15 cm square up to 1 or 2 m; as the tank size increases, so does the price. They all work on much the same principle. Transducers fixed to the bottom of the tank are activated by an ultrasonic oscillator and the waves produced are transmitted through the water in the tank. The cleaning action is by cavitation, i.e. the formation and implosion of vast numbers of tiny bubbles in any crack or cavity. They are useful for cleaning robust fossils covered with a matrix which is sufficiently pervious to be wetted with water containing a small amount of a wetting agent or detergent. It is usual to place the specimens in beakers of fluid and stand these in the water in the tank. With care and practice the apparatus can be used to develop micro-fossils by placing these in a suitable aqueous solution in a test tube and dipping it into the water in the tank. The length of time the tube is in the agitated water can be very critical. It was found that with certain ostracods, five seconds treatment would clean and separate the shells but seven seconds smashed the specimen to pieces. Small cup corals have been developed in this way. When using such a tank it is necessary to remember that the action takes place in any cavity, so badly cracked specimens are at risk. The frequencies at which these machines operate vary considerably, and they all produce an audible noise. The noises produced by some are intolerable to certain people, yet can be ignored by others. The makers issue warnings concerning the types of liquids which are not safe to use in these machines. These warnings should be heeded as they do not exaggerate the dangers which can arise if certain solvents are employed, even in test-tube quantities.

The second form of ultrasonic equipment available is the probe described by Macadie (1967). This is a probe fitted with a pistol grip and activated by an ultrasonic oscillator. It is constructed so that ultrasonic waves can be directed at one point on a specimen immersed in a vessel of water. Macadie used it successfully to remove the insoluble residue left on certain fossil fish skeletons after development in acid.

The third type are the dental machines made for the ultrasonic treatment of teeth. Hempel (1969) described the application of one of these, the 'Cavitron', for cleaning statuary. This apparatus has also been found to be of great use in the development of small specimens in palaeontology, as well as fossils as large as the limb bones of a plesiosaur. The 'Cavitron' is also activated by an oscillator and the way in which the waves are made to operate the cutting tool is highly ingenious and allows the handpiece to be small enough to be held like a pen.

Two types of handpiece are available, one in which the tools are cooled externally by a jet of water, the other with internal cooling. External cooling sets up a fine mist thus making work under magnification difficult, whereas the other handpiece allows the specimen to be worked dry. It is not necessary to connect the water feed of the apparatus to a mains water tap; satisfactory operation is achieved if the input tube is placed as a syphon into a suitable reservoir (e.g. a 5l polythene bottle) about 1m above the apparatus. With this arrangment the apparatus may be placed on a trolley and used anywhere in the laboratory where there is an electric outlet socket. When the externally cooled handpiece is used, the waste water is caught in a shallow dish.

No ultrasonic apparatus is cheap, and if a laboratory is not rich enough to have all of the types described, the decision as to which to buy will be dictated largely by the type of specimens most commonly treated.

Most manufacturers will arrange demonstrations, and very often apparatus can be seen at exhibitions of scientific and related apparatus. All of the tools so far described may be used to develop a small fossil completely or, at least, may be used in part of the work.

The Desoutter Pneumatic Power Pen Type V.P.2

In recent years a tool has been added to the preparator's armoury which is too powerful for use on small specimens but is invaluable for larger ones. This is primarily intended for use in industry and was first employed in palaeontology by Jones (1969) to prepare the remains of a Middle Jurassic dinosaur preserved in a series of

Plate 7. Pneumatic power pen V.P.2. (By courtesy of Desoutter Bros. Ltd.)

ironstone nodules. The tool is capable of dealing with matrices much too tough for a Speed Engraver; in fact it can do anything which a hammer and chisel can do, and very much more quickly; it is very light and easy to use.

It is a small pneumatic vibrator requiring a pressure of only 40 lb/sq. inch (2.8 kg/cm^2) and can be operated successfully if powered by an air compressor made for use with a paint spray. A combined filter and mist oiling attachment is available as an extra.

The point is tungsten-tipped and very hard-wearing and, by a fortunate coincidence, is ground at an angle which splits the matrix very cleanly off the specimen; the only disadvantage is that the point is rather short and is difficult to use in steep-sided narrow depressions. A ganoid fish in a hard clay nodule containing much pyrites which was difficult to develop by hammer and chisel, was very quickly cleared of matrix with the power pen. The tool is illustrated in Plate 7. I am indebted to Messrs Desoutter Bros. both for the photograph and their permission to publish it.

Fossils in Soft Matrices

Many fossils are not preserved in hard rock but in loosely consolidated matrices. Those found in some sandy soils and clays come into this category and no refined tools are required to develop them. Often they can be cleaned, either wet or dry, by brushes of various kinds. Ordinary tooth brushes are good, but nearly all those currently made have handles which are soluble in most laboratory solvents, so should be restricted to dry use or use with water. The brushes made for cleaning typewriters are, as a rule, fitted with wooden handles and these can be used with organic solvents. Dental brushes, made for use in a rotary handpiece, can be used either with the handpiece or mounted in a pin vice as a hand tool. The small brushes made for polishing teeth are particularly useful for small specimens or the removal of matrix in deep depressions or ornament on larger ones. A chuck called a 'porte polissier' can be bought to hold these. Mounted in the end of a piece of wooden dowelling of small diameter, or a piece of polythene rod, it can be used as a hand tool or, without the mount, in a rotary handpiece. The brushes are available in three grades – hard, medium and soft.

Some vertebrate fossils collected from desert sands are so friable that it is necessary to consolidate them before all the sand has been removed from the surface of the bone. After the specimen has been sufficiently strengthened, the residual sand can be cleaned off by the careful use of brushes wetted with the appropriate solvents. It occasionally helps if the solvent is mixed with a non-solvent, such as water; the effect of this is to stop the sand hardening up again. This mixture should not be used close to the bone surface, and here only enough pure solvent should be

Comminuted bones of giant marine reptiles 81

utilised to soften the surface of the plastic holding the grains to the bone. At close quarters to the bone, removal of sand grains can sometimes be achieved by gentle scraping with a piece of soft wood sharpened to a chisel point.

The opposite of this condition occurs in some fossils both vertebrate and invertebrate found in clays of the type commonly found in excavations for brick-making. Here, some of the fossils are so robust that they can be put under a running tap and scrubbed clean, but before so violent a treatment is commenced it is better to do a test washing of a small area in still water to establish that no harm will come to the specimen. The specimen should be checked to see that it has not already been broken and held together with hardened mud. In this case it would certainly have to be taken to pieces and repaired with something stronger than mud, but this is better done under control in a bath of still water, or by softening the mud in one break at a time with water applied locally by brush and picking it out with a needle.

The Unpacking and Preparation of Comminuted Girdle Bones of Giant Marine Reptiles

A special technique for collecting these large, and very often, badly shattered fossils has been described in chapter 3. The methods employed to unpack and develop them differ so greatly from those used on other fossils that they must be considered separately.

The procedure is first to remove the last layers of plaster bandages applied in the field; thus revealing the underside of the specimen. All the matrix, except that filling cracks in the bone, is cleaned off and the bone consolidated where necessary with a solution of Butvar B98 in isopropyl alcohol. If the bone is so comminuted that there is little or no hope of correcting any distortion by taking the whole thing to pieces and re-assembling it, matrix should now be picked out of the wider cracks with a needle and these filled with a gap-filling adhesive of the jute floc-based type. This can be done with confidence, for the pieces are held in position by the glass mat and Butvar B76 on the other side.

These bones are often so thin that they must be kept on a permanent support. It is better that this should be removable, so that both sides are available for study. The support can be made either of plaster of Paris or fibreglass, the latter is preferable.

If plaster is used, the side just treated is given a coating of soft soap solution; over this wet paper is applied. A dam of plasticine or clay is built to the required height and the plaster poured in; the top, which on turning will be the bottom, is flattened off.

When fibreglass is used it is essential that the resin does not come into direct contact with the bone. Should it do so, the support will be irrevocably stuck to the specimen. This may be avoided in several ways. Provided that the surface of the bone is stable enough to allow the removal of rubber when it is dry, without damage, three or four coats of rubber latex can be applied and when dry the upper surface lubricated with silicone oil or grease. Alternatively, a thin sheet of rubber can be made by pouring latex onto a piece of plate glass and allowing it to dry. Before removing the dry sheet from the glass it is necessary to dust the surface liberally with french chalk or talc and to apply the same powder to the underside as it is peeled off. Failure to do this results in the rubber sticking to itself and becoming a useless mess. This sheet can then be draped over the specimen and the upper surface lubricated. In place of rubber, a sheet of tough plastic foil like 'Melinex', which is made of terylene, may be used. Thin polythene sheet will serve the same purpose, but whatever the foil, it must have no holes in it and must be strong enough to remain unbroken while the glass mat and resin are applied. Substances other than rubber require no lubrication on the side which receives the glass mat, but as an added precaution against punctures it is as well to give the bone a light coating of olive oil. This cannot be used if the film is rubber. The covering film may be held in place by pieces of 'Sellotape' passing from it to the jacket on the other side of the bone.

Chopped strand glass mat is cut into suitable lengths and shapes and applied to the covered specimen, using a low viscosity polyester resin made for this purpose with the recommended percentages of accelerator and catalyst. The resin is stippled into the mat with a stiff brush and each piece of glass mat applied must overlap its neighbour. The number of layers of mat required varies with the size and weight of the specimen, stiffening ridges may be formed over paper rope which can be bought in different thicknesses. When the resin has set the support is removed and the edges trimmed by grinding, which should be done under a dust extraction hood; the operator should use a mask and protect his hands and forearms with gloves and industrial sleeves.

Now that the underside has been prepared, the specimen on its support can be turned over. The plaster bandages where present, are removed and the glass mat put on in the field is taken off. This is done by dissolving the Butvar B76 in acetone, the removal can be speeded up by placing pads soaked in the solvent on top of the strips, covering them with polythene sheet weighted down at the edges and leaving them in place for some time; this keeps the acetone in contact with the plastic and stops evaporation. It is safer if small areas of field coating are removed gradually rather than to attempt the removal of the glass in large sections at a time. Butvar B98 is not soluble in acetone, so the bone treated with this is not affected. When all the glass mat has been removed, preparation of the top side can be completed.

References

Hempel, K. F. B. (1969) 'The restoration of two marble statues by Antonio Corradini', *Studies in Conservation*, **14,** 126–31.

Jones, M. D. (1969) 'A pneumatic power-tool for the palaeontologist', *Mus. Assist. Group*, Newsletter, March.

Macadie, C. I. (1967) 'Ultra-sonic probes in palaeontology', *Journ. Linn. Soc.*, **47,** 251–3.

Newman, B. H. (1968) 'The jurassic dinosaur *Scelidosaurus*, Owen', *Palaeontology*, **11,** 40–3, pls. 7 and 8.

Rixon, A. E. (1963) 'The use of industrial diamonds in palaeontology', *Indust. Diamond Review*, **23,** 123–5.

Stucker, G. (1961) 'Salvaging fossils by jet', *Curator*, IV, 332–40.

5
Chemical Methods of Development

The use of chemicals for the development of invertebrate fossils preserved as silica, carbon or chitin is an old technique, going back at least as far as the last half of the nineteenth century. Until the late 1930s, chemical methods were not used to any extent to prepare vertebrate remains, at which time Mr Toombs was assisting Dr E. I. White in the British Museum (Natural History) with his work on the ostracoderms from the Old Red Sandstone. One of their problems was that the ornamented side of the boney plates always remained in the rock when it was split mechanically and only the smooth inner surface of the specimen was revealed. Since indentification of the specimen depended on the pattern of the ornament some way had to be found to clear this of rock; this led Toombs to experiment with various acids to find one which would facilitate the removal of the matrix from this side. He discovered that a dilute solution of acetic acid in water would do this and published his results in 1948.

Acetic Acid in the Preparation of Vertebrate Remains

The theory on which this method is based is that while dilute acetic acid will dissolve the carbonates in the matrix, it does not significantly effect the apatites which make up the greater part of fossilised bone. The acid must be diluted for it will not react unless it is ionised. Toombs (1948) used a dilution of the acid in water of 33 per cent (v/v), but later work showed that at this concentration a large part of the acid is wasted, and that concentrations of between 10 per cent and 15 per cent are more efficient for most purposes. In cases where the rock contains a great deal of carbonate, concentrations as low as 1 per cent of the acid are advisable in order to reduce the violence of effervescence due to the expelled carbon dioxide, which can damage areas of bone in which the walls are thin.

The removal of the carbonates from the cells and cavities in the bone weaken it and some artificial means of restoring its strength must be used. When dealing with the more or less flat plates of ostracoderms Toombs first used 'Bostick' paste – a rubber compound. He applied this to the exposed unornamented side before placing the specimen in the acid solution. It was not the ideal answer because, after preparation, one side of the specimen remained covered with an opaque substance. This did not matter a great deal in the case of the ostracoderms, but the extension of such a potentially valuable method to the development of whole fish skeletons was desirable and, in such specimens, it was of importance that both sides of the bone should be seen.

The advent of cold-setting transparent polyester resins made this possible.

The Transfer Method

Toombs and Rixon (1950) described the transfer of fossil fish skeletons from their matrix to blocks of transparent polyester resins.

When a slab of matrix containing a fossil fish skeleton is split open, the bones usually split through their centres and what is seen is a longitudinal section through the vertebral column, the skull, and fin rays. This is not an ideal state of affairs since it is only possible to see the anatomy of the inside of the skull and vertebrae, and the exact identification of the lateral fins is very often not easy and is sometimes impossible.

It was therefore desirable that a method should be found whereby the matrix-covered lateral aspects of the bones could be revealed without obscuring what was already visible and at the same time the relative position of the bones should remain unchanged. Very often, parts of the skeleton are preserved in the form of natural moulds in the rock and it is of great importance that these should not be destroyed. All of these conditions can be satisfied in the following way.

The rock containing the specimen is ground so that the sides are rectangular and as much is ground off the bottom as is safe. This operation is best carried out on a wet bandfacing machine; if the specimen is too weak, this treatment may be omitted.

Chemical Methods of Development

A container is made by folding light gauge polished aluminium sheet (Fig. 7). This should be large enough to contain the specimen with at least 1 cm to spare at each side and have walls twice the height of the specimen. The piece of rock is placed in the box, fossil side up, and positioned so that its sides are equidistant from the sides of the container. If, for some reason, the block of rock has to be more than about 2 or 3 cm thick, molten polyethylene glycol 4000, or, failing that paraffin wax, may be poured into the container so that the excess matrix is buried in it when it has cooled. This is a method of saving resin and is not essential to the technique.

A low percentage of accelerator is stirred into a quantity of polyester resin as near water-white in the polymerised state as possible. The percentage will depend on the resin used and this mixture may be made in bulk as it has a long shelf life if kept in a cool place. The British resin Trylon E. M. 301 requires to be mixed with 1 per cent of Accelerator E.

An estimate of the amount of resin required to form a layer about 6 mm ($\frac{1}{4}''$) thick from the bottom of the container and extending up the sides of the block is made. This amount of pre-accelerated resin is mixed with 2 per cent of catalyst, which may

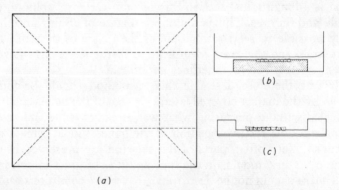

Fig. 7. Stages in the transfer process. (*a*) Aluminium sheet marked out for bending; (*b*) Section through tray and plastic block containing the specimen; (*c*) Section through the plastic block after preparation, showing the cavity left by the dissolved matrix, the newly exposed bones and the glass plate applied to the outer surface of the plastic.

be either in the form of a paste or a liquid. A liquid catalyst may be measured by volume, but care is needed as all such liquids are organic peroxides. The mixing must be thorough. The use of paper cups reduces time spent on cleaning apparatus. The resin mixture is poured into the container and allowed to set. At normal room temperature this usually takes from four to six hours. The setting time is made long deliberately as the reaction is exothermic and the slower the resin sets, the less heat is generated at any given instant and the less likely is it to crack. For the same reason, the depth of resin poured at any time is restricted to 6 mm ($\frac{1}{4}''$). The pouring of resin in such layers is continued until the specimen is covered to a depth of about 6 mm above the level of the skeleton.

When the last layer of resin has fully set (which usually takes 4–5 days), the block with the enclosed specimen is removed from the container by unfolding the metal sheet and if wax has been used it is cut away. The bottom of the block is ground on the bandfacer to remove any resin which may have found its way onto it, and to expose the rock; the sides are trimmed square and the resin surface, below which the skeleton lies, is carefully ground flat. A piece of window glass is cut so that it is slightly less in breadth and length than the resin block. When the block has been thoroughly dried some resin is mixed and a thin layer poured on the face above the specimen. The glass is lowered onto the unset resin in the same way as a cover slip is applied to a microscope slide. One shorter side is placed on the surface adjacent to a similar side of the block, the sheet of glass forming an angle with the face of the block, and the angle closed slowly. Any air bubbles are expelled by pressing on the glass. When the resin sets the skeleton can be seen clearly through the glass. Any excess resin which has flowed onto it can be scraped off with a razor blade. The specimen is now ready for acid treatment.

A polythene bowl is partially filled with a 10–15 per cent (v/v) solution of glacial acetic acid. There is no need for accurate measurement of the acid and water, as a visual estimation of 1:9 or 1:6 is quite easy in a straight-sided vessel and there are wide tolerances. The block of resin containing the specimen is placed in the acid with the exposed rock uppermost. Reaction between the acid and the carbonate in the rock will begin at once and the speed of solution can be judged by the bubbles of carbon dioxide which are given off. From time to time it is necessary to remove the insoluble residue, as this tends to slow up the solution of the rock below. While there is still a considerable layer of matrix above the specimen the insoluble residue may be brushed off, but when the level of the skeleton is approached the safest way of removing the debris is to take a polythene pipette, dip the end into the acid well away from the specimen and fill it with liquid, then without removing it from the acid, move it over to the specimen, and holding it just above the surface of the rock, gently squeeze the bulb. This will cause a movement of the liquid above the rock and the loose undissolved particles will be washed away. It may be necessary to change the acid solution from time to time as it becomes exhausted. A specimen should never be left long in acid which has become heavily charged with salts, and certainly no extensive evaporation must be allowed to occur as the salts will crystallise and may damage the specimen.

When bone begins to show, the specimen is removed from acid and washed in water to remove excess acid and all salts. The washing can either be done in slowly running water or constant changes of very large volumes of still water. The washing completed, the specimen is dried in air. It is not wise to dry it in an oven, unless this is of a type which can be set at a temperature only a few degrees above the ambient, for excessive heating can cause the plastic to crack. The hot air stream from a hair drier may be used, provided that none of the bones are loose.

The dry specimen is examined under a binocular microscope and if the bone is covered with loose particles of undissolved matrix, these can be removed by the careful use of a small moistened paint brush. On more robust bone, a fine needle ground to a chisel point may be used. The exposed bone can, if necessary, be strengthened by using a weak solution of polybutylmethacrylate in methyl ethyl ketone. Some workers use a solution of polymethylmethacrylate and some a proprietary substance called 'Glyptal', but having tried these and others, the author still prefers polybutylmethacrylate since it is practically foolproof. When the plastic solution has thoroughly dried, the specimen is returned to clean acid solution and the entire process repeated until the whole of the bone has been exposed.

If the insoluble fraction of the matrix is very difficult to remove,

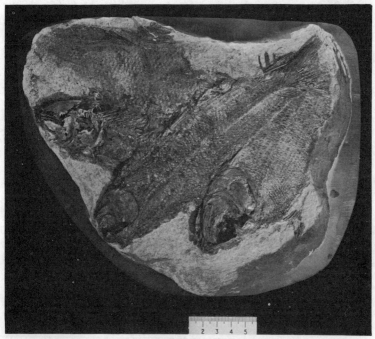

Plate 8. *Tharrhias*. Developed with dilute acetic acid, using the 'Transfer Method'. This aspect of these specimens was invisible before preparation. (By permission of The Trustees of The British Museum (Natural History).)

the ultrasonic probe described by Macadie (1967) helps, or a Cavitron may be used; the latter is less cumbersome.

The completed preparation will show all the parts of the skeleton previously covered and any which were preserved as natural moulds will be reproduced as faithful casts in the resin. The specimen is now visible from both sides and the cavity left by removal of the rock results in it being protected on all sides but one. Protection for this side can be provided by a piece of cardboard held in place by rubber bands. Specimens of this type are easy to examine, easy to store safely and easy to pack for transport.

Objections have been raised to this method on the score that nobody knows how long the plastic will last. However, if the plastic is used properly, with regard to the quantities of accelerator and catalyst employed, the earliest types are known to be capable of lasting longer than twenty years, since many as old and older than this are still in good condition. With the improvement in resins the chance of survival should increase. In any case, good photographs and drawings can be made and the information thus obtained cannot be gathered in any other way. A good cast of the specimen in its original state preserves all the data available before preparation.

It is curious that no objection seems to be raised to the practice of serial sectioning specimens, even though by this method the object as a whole is never seen except as a reconstructed model or drawing. This is not to be taken as a condemnation of the method for its use in almost every branch of palaeontology has provided information of the greatest importance, and science owes a great debt to those who first developed the technique and to those who continue to improve and make more accurate the apparatus used. It is the only way in which some specimens can be studied. This argument fully justifies the serial sectioning technique and it equally justifies the transfer technique and all other chemical methods.

The Development of Fossil Vertebrates in the Round by Acetic Acid

So far the specimens discussed have been essentially flat or of a type in which the disarticulation of the skeleton was not desirable. In dealing with fossil reptiles and mammals the preparator is normally required to expose all sides of the bone so that they can be studied from all aspects and the disarticulation of jaws and other joints is usually desirable. This cannot be done if they are attached to blocks of plastic. For preparative purposes, these specimens fall into the following groups:

(1) Isolated bones
(2) Isolated skulls, with or without jaws attached
(3) Isolated small mandibular rami
(4) Isolated but articulated joints, ankle, wrist, knee, etc
(5) Complete or partial skeletons
(6) Cave breccias, fissure fillings and bone beds.

Group 1: Isolated Bones

Unless these are badly distorted or crushed they present very little difficulty. If the physical condition will permit, the amount of matrix covering the bone may be reduced by mechanical methods; this speeds up the preparation. Any exposed bone is coated with a thin solution of polybutylmethacrylate in methyl ethyl ketone, particular care being taken with comminuted or broken areas. The plastic, however, must be allowed to dry out completely before the specimen is put into the acid and care should be taken not to build up too thick a coat. The bone should be placed on some form of grid or perforated tray which will not be affected by the acid. The diffusion grids used in some types of strip lighting make very good supports, or special trays may be made from glass fibre (see Appendix I).

The specimen on its support is placed in the 10 per cent acetic acid solution and left until about 3 mm of matrix has been dissolved or softened. This can be judged by gently brushing an exposed area of bone while it is still in the acid. As a rule the newly exposed bone is lighter in colour and the extent of light-coloured bone revealed is a measure of the rate of solution of the rock.

The specimen on its tray is now lifted out and placed in a large vessel of water. The water should be changed every eight hours and the bone should wash for at least three times as long as it has been in the acid. After washing, the specimen is dried at 60°C in an oven. If a plastic, other than fibreglass is used as a tray, it should first be checked to see that it does not soften at this temperature. The dry specimen is examined and if it has crystals of salts on it, it should be cooled, rewashed and redried. The dry specimen is now cleared of softened matrix by fine brushes and needles and the newly exposed bone treated with the polybutylmethacrylate solution.

All this work must be done under a binocular microscope since structures are often revealed which would be lost in mechanical development. This is most likely to occur when the joint surfaces of reptile bones are treated for sometimes a thin layer of bone which is deeply pitted is seen, and unless a microscope is used, it can easily be mistaken for encrustation of polyzoa and removed in error. The specimen is finally returned to the acid bath and the cycle of treatment is repeated until the bone is absolutely clear of rock.

The appearance of the finished bone can be improved if most of the surface polybutylmethacrylate is removed; this can be done with a soft brush and solvent. The solvent must be clean and needs changing after the brush has been dipped into it a few times and should therefore be used from a small vessel. Small nickel evaporating dishes are excellent for this purpose. During cleaning with solvent there is a danger that the fingers may become stuck to the specimen. If this should happen do not attempt to separate them by force. A brush moistened with the solvent and drawn between the skin and the specimen will free it without damage. When as much as possible of the surface coating has been removed, the bone is allowed to dry and is then placed in a 1:3 mixture by volume of polyvinyl acetate emulsion and de-ionised water. Vacuum impregnation may be used or, in the absence of suitable apparatus, the specimen may remain in the emulsion for several hours. On removal from the mixture the excess is allowed to drain off and any surface pools or 'tears' are removed with a soft brush. It can be left to dry in air on a piece of paper provided that it is moved at intervals, or can be stood on a zig-zag of metal such as has been previously described (see Chapter 2). If, when the bone is dry, it is not considered to be strong enough, the impregnation may be repeated. Provided that the treatment has been carried out properly the finished specimen should be strong and have a natural appearance.

The filling of wide cracks and the restoration of missing pieces may take place at any stage in the treatment, according to the circumstances. Where an area of bone is badly crushed and comminuted and there are wide gaps between the pieces, the filling should be done either before the treatment with acid starts, or as soon as enough matrix has been cleared from the gaps to allow the insertion of a filling compound. In some cases, where the filled specimen has to be returned to acid, a polymethylmethacrylate monomer/polymer mixture (North Hill Plastics, etc;) should be used. When a thin layer of bone is separated from the main body, such as sometimes occurs on the joint surfaces of reptile bones, a thin mix of this material may be inserted between the two layers of bone, using a polythene pipette with a fine point.

Development by acetic acid

Sometimes it is an advantage not to fill cracks in the bone until the job is nearly completed, because they give access to the interior and through them the inside of the bone may be strengthened with polybutylmethacrylate solution. Such cracks can be filled before the final removal of the surface plastic, with plaster of Paris. When tinted by the addition of a little yellow ochre powder before mixing, the hardened material contrasts with the colour of the bone and gives a more pleasing appearance to the finished specimen than does the glaring white of pure plaster.

Group 2: Isolated Skulls with or without Lower Jaws Attached

Skulls from about 7.5 cm to 20.0 cm in length, which are uncrushed, are not distorted and have no lower jaws attached, are not very much more difficult to deal with than an isolated vertebra. There are, however, one or two weak points in a skull and their existence must be taken into account.

Teeth, especially the long canines of some mammals, and all reptile teeth are easily damaged; for this reason skulls should always be placed in the acid solution so that the weight is taken on the top. The nasal area becomes weak when the matrix has been removed from the nares, hence the acid attack in this area should be inhibited until the last stage of the preparation when all bones on the outside of the cavity have been strengthened. This can be done by coating the matrix in this area with a thick solution of polybutylmethacrylate. In some cases it may be necessary to augment this by a coat of rubber latex; the foramen magnum may be treated in the same way. When the point in preparation is reached at which the solution of the matrix in these areas is required, the plastic can be scraped away with needles or dental tools.

The general procedure for dealing with skulls is much as has been stated above for isolated bones. All visible bone is coated with polybutylmethacrylate and allowed to dry. The specimen is placed in a tray or on a grid with the tooth side uppermost. It may be retained in this position by cylinders made by rolling up cotton or linen rags and tying them with thin string. These are soaked in water and packed round the skull and the whole placed in the acid. When about 3 mm of matrix has been softened the skull is washed and dried. The loosened matrix is removed under a binocular microscope, using the same tools as before or in some cases a dental scraper. The specimen should be held lightly in one hand or, if it is too large for this to be done safely, it can be supported on a laboratory jack, the top of which has been covered with some sort of soft material, and steadied with the hand. When a laboratory jack is used, never let go of the skull; if you do, it will almost certainly fall off.

The matrix should first of all be removed from the teeth and newly-exposed areas strengthened with polybutylmethacrylate. The coating should not be thick, as palaeontologists are becoming more and more interested in tooth wear and over-coating can obscure wear facets. A certain amount of commonsense must be used however, and if a tooth is very badly cracked and it becomes a question of obscuring the wear facets or losing the tooth altogether, it is better to risk the ire of some workers and at least preserve the gross anatomy of the specimen.

The palate and associated areas are next dealt with. When the underside of the skull is as strong as possible, work is carried out upon the sides and the top. The most dangerous phase of the operation is when the top is being cleaned, for then the delicate teeth and palate areas are more at risk and their welfare should always be borne in mind. This is why, whenever possible, the specimen should be held in the hand as the sense of touch will give warning of a potentially dangerous situation before it actually develops.

Skulls of 5 cm (2") or less in length are not very easy to hold in the hand safely during preparation. This type of specimen is best supported in a block of polyethylene glycol 4000 while needles and brushes are in use. The block of soluble wax should be formed on a piece of hardboard, for some specimens of this type are better not heated in ovens to the temperature required to remove the wax. In these cases removal is carried out by solution of the wax in warm water; hardboard is used because it rapidly becomes waterlogged and sinks, so there is no fear of the block reversing itself and the specimen falling to the bottom of the vessel. The temperature of the water can be about 40°C and the wax may be recovered after the removal of the specimen by evaporating off the water. The solution is poured into a flat-bottomed dish which is very much greater in volume than liquid; evaporation at room temperature is

Development by acetic acid 95

quite rapid and no heating is required.

The final treatment of the skull is the same as for an isolated bone but more care is needed. The thinned emulsion is best applied by brush in this case, to ensure the control of the amount deposited on the teeth.

When the skull has the lower jaws attached, the procedure depends on the amount of matrix between them and the palate and the rate at which the rock dissolves in the acid. If, either as an accident of fossilisation or as a result of previous mechanical development, there is little matrix in this area and the rock slowly and only partially dissolves, the preparation follows the same lines as for a skull without lower jaws. This is commonly the condition of reptile skulls from parts of the Karroo. The gradual removal of the matrix and strengthening of newly exposed bone is carried out on all sides, but particular attention is given to the areas between the upper and lower tooth rows and between the articulars and the quadrates.

Work on the lingual surfaces of the lower teeth is difficult especially if the angle between the rami is acute. On the larger specimens, dental mirrors and the light guide described by Macadie (1966) help to make these areas visible; in the case of specimens in which there is not enough room to use them and a needle, small pieces of mirrorised Melinex attached to thin flexible wires are useful. It is nearly always necessary to work in these places with bent needles and sometimes the metal shafts of small paint brushes must be bent so that polybutylmethacrylate solution may be applied to the teeth.

Reptile teeth are often very fragile before they have been strengthened so a minimum of force should be used to remove the softened matrix. Fine brushes should be used whenever possible and splinters of bamboo or pieces cut from the shafts of small feathers are often preferable to metal needles. The space between the upper and lower tooth rows is sometimes very narrow, and the same is true of the gap between the articular and quadrate in reptile skulls. When it is necessary to use needles in these areas they should be ground as flat as possible or fine gauge nichrome wire may be hammered flat and used in place of steel needles.

When the matrix has been cleared from between the greater part of the upper and lower tooth rows, a small excess of

Chemical Methods of Development

polybutylmethacrylate should be introduced at several points. The object of this is to hold the two rows together lightly so that they cannot move before the preparator wants them to. A point is reached when the two tooth rows have no matrix between them and, as far as can be seen, the quadrates and articulars are free from each other and the jaws are only held in place by the excess plastic. This is then removed, as far as possible, by local application of methyl ethyl ketone and then the whole area between the tooth rows and the quadrates and articulars is flooded with solvent. The rami are then lightly gripped at the centres and gentle traction is applied so that if, there is movement, the jaws will hinge on their natural joints. When the jaws do move, the softening of the plastic and gentle traction is continued until they separate from the skull. If there is no movement, no great force should be exerted, but it should be assumed that there is still matrix in the articular quadrate area. The specimen is dried and the joint zone gently scraped with a needle to break the plastic film present on the inside and the skull is finally put back in acid. This cycle is repeated until the joint is free and the jaws can be removed.

All of this may sound a hair-raising and drastic business. The first attempt is rather nerve-racking, but this is now a proven technique which is within the capacity of any careful preparator. The operation has been carried out by the present writer and several other workers on dozens of specimens and, as far as is known, no disasters have occurred.

There are matrices, in which vertebrates are found, which dissolve quite rapidly and unevenly in acetic acid. The rates of solution in different areas of the same block can vary dramatically. It is quite clear that to remove the jaws from specimens enclosed in these rocks, it is unsatisfactory to put them into acid and hope that the matrix between the tooth rows and the articulation of the jaws will dissove faster than that in other parts of the specimen. Some method has to be employed to confine the acid attack to the area in which it is wanted.

This can be done if all the specimen except the area from which the removal of matrix is required is first enclosed in an acid-proof casing. Rubber is impervious to and is not attacked by acetic acid, so this can be used as a base for the protective coat. But rubber is flexible and, as has already been said, the matrix can be expected to dissolve unevenly which may produce areas of weakness extending from the unprotected side into the rock; and if the specimen is held in a flexible container, fractures may occur through these areas. The casing must therefore be made rigid. The method detailed below is used to fulfil these requirements.

(*a*) All exposed bone is coated with polybutylmethacrylate solution which is allowed to dry.

(*b*) The whole of the top and sides of the skull, down to a line about 3 mm above the alveoli of the upper teeth is given a coat of silicone fluid followed by two coats of rubber latex, the first being allowed to dry before the second is applied. The latex is applied to both bone and matrix. If the orbits, foramen magnum or any other natural cavities are deeply excavated, either by nature or by previous development, they should be packed with paper soaked in water containing a little ammonia. While wet, these areas are painted over with rubber latex and the coat extended to cover about 6 mm ($\frac{1}{4}$ in) of the dried latex surrounding the cavity. This rubber is allowed to dry and a second coat applied.

(*c*) Some short staple jute or cotton floc is damped with a 5 per cent solution of 0.88 S.G. ammonium hydroxide solution, and is mixed to a spreadable paste with rubber latex. A layer of this paste (3–6 mm thick) is spread over the set latex on the specimen. The specimen is then set aside for this coat to dry, which at ordinary room temperature will happen overnight. The skull, apart from the lower jaws and the ventral aspect is now enclosed in a watertight case.

(*d*) Take 70g of polyethylene glycol 4000
        23 g of glycerol
        15 cc of water

Mix cold, then warm slightly and work into a smooth paste. Allow to cool and then mix in a little at a time:

        29 g of precipitated chalk for every 100 g of mixture.

The result will be a white putty-like substance which is soluble in water.

Apply this to the teeth and to the adjacent bone. Cover the rest of the ventral side of the skull with tissue paper. Support it, ventral side down, in a sand tray or in some other way.

(*e*). To the rubber apply a gel coat of a polyester resin. This is thixotropic and will not run off the vertical surfaces. The resin chosen should be one with high resistance to chemicals and the formula for the gel coat will vary with the brand. If the British Trylon is used the formula is:

    Trylon Gel Coat Resin. G.C. 150 P.A.   100 g
    Liquid catalyst   2 g

This resin has the accelerator already mixed into it by the manufacturers. The coat is applied by brush which may be cleaned first in acetone and then in a strong solution of a detergent, followed by a wash in water.

(f) Chopped strand glass mat is cut into strips about 4 cm wide and into lengths, some of which will span the length and some the breadth of the rubber covered area. These are stippled onto the rubber, so that each slightly overlaps the other, with a stiff brush loaded with polyester resin, accelerator and catalyst mixture. For Trylon the formula is:

    Trylon Polyester Resin. W.R. 180   100 g
    Accelerator   2 g
    Liquid catalyst   2 g (2 cc)

A bridled glue brush is the best type for this purpose. Each strip of glass should be fully impregnated with the resin, but care is needed to avoid excess running off the pendant ends. The object of the application of the water-soluble putty in stage d is to stop resin getting onto or between any exposed teeth. When the glass fibre strips have been applied to both length and breadth and the resin has set the specimen is in a watertight rigid box. Rough edges may be trimmed off with a diamond dental saw.

(g) The water-soluble putty is removed, most of it by just pulling it off. What is left may be washed off with warm water after the paper has been removed.

The specimen is now ready for acid treatment. This is carried out in the usual cycle of solution, washing, cleaning of newly exposed areas and strengthening the bone until the two tooth rows are free from each other and the jaw joint is free from matrix. The jaws are then separated as described above. At this point both the upper and lower teeth are finally cleaned and strengthened with polybutylmethacrylate solution. The lower jaw can now be worked separately.

The further development of the skull makes it necessary to

remove the fibreglass and rubber gradually from all areas. The removal is done by cutting through the fibreglass with a rotary diamond dental saw and the rubber with a sharp scalpel and scissors. The areas to be uncovered first will depend on the specimen and the preparator must exercise his own judgement in this matter.

By this method, the skulls of ichthyosaurs have been completely disarticulated, so that the bones can be studied individually in the same way as those in the skulls of recent animals. If the job is done with care, bones no thicker than paper can be extracted without damage and the sutures kept intact.

Group 3: Isolated Small Mandibular Rami

The mandibular rami of both mammals and reptiles often survive when the skull has been completely destroyed, either before or during the process of fossilisation. They may be less than 2.5 cm in length and in many the whole tooth row is preserved. The cusps of the teeth may be no larger than a small needle point. Care is therefore needed during preparation. When the specimen is on a large piece of matrix it should be cut out, using a rotary dental diamond saw, on a block, the edges of which are at least 6 mm from the specimen on all sides. No attempt should be made to remove matrix by mechanical means from the specimen, and especially not from the teeth. The back and sides of the block of matrix are given two coats of rubber latex, so that the acid attack is limited to the top. A small piece of waste matrix is placed into a 10 per cent (v/v) solution of acetic acid and the speed and violence of the reaction is checked. Should the reaction be too violent a 1 per cent solution should be used in the actual preparation. All exposed areas of the specimen are coated with polybutylmethacrylate solution and this allowed to dry thoroughly. The block containing the specimen is supported on a narrow meshed grid with the specimen uppermost. The grid may be made of a piece of discarded nylon stocking fixed to a suitable frame; a polymethylmethacrylate frame is excellent and the piece of stocking can be fixed to it with a rubber band. The whole is now lowered into the acid solution.

A constant watch should be kept on fossils of this type while they are in acid, so that if anything appears to be going wrong the specimen can be rescued before any damage is done. From time to time the undissolved residue of the matrix should be gently washed away with a polythene pipette, particular attention being given to the edges of the specimen and the teeth. If the teeth were not exposed before preparation started, as soon as they appear the specimen should be lifted on its grid from the acid and washed in several changes of water. No solid matter either in the acid, or on the grid or in the washing waters, should be thrown away at this stage. Gentle heat may be used to dry the specimen, but if polymethylmethacrylate frames are employed this must be well below 80°C, as this substance has a shape memory and the frames will revert to straight pieces and damage to the specimen is almost certain.

The jaw is now examined under the binocular microscope, and softened matrix on the block around it is removed with needles; that on the bone and teeth also may be removed with needles if the structures are strong. If there are any cracks in the bone, and certainly if the teeth show signs of damage, it is better to use a splinter of bamboo or a section of the shaft of a small feather to remove the loosened matrix, although with care a really skilful worker can use a specially ground steel needle.

The teeth of small animals often show wear facets and, as previously mentioned, these are of great interest to vertebrate palaeontologists. If the preparator is not familiar with the appearance of such markings he should consult somebody who is before he starts applying polybutylmethacrylate to the teeth, as an over-application can obscure these and it is easier and safer to limit the amount of plastic at this stage than it is to clean it off later. Great care is needed when applying the plastic solution to cracked teeth, as small pieces may be dragged into the brush hairs by capillary attraction. Sometimes it is safer not to use a brush but to cut out and shape to a point part of the vane of a small feather and use this instead. The piece of vane may be fixed to any convenient handle; a small pin vice does very well. The shed feathers of cage birds, such as budgerigars, are good for this purpose since their colour greatly contrasts with that of a piece of tooth and any part which should by mischance become stuck to them can be seen easily and recovered. The important thing is not to let a piece become detached from the tooth. On fresh material this is usually

Development by acetic acid

easy, but specimens from old collections which have been developed manually often have cusps broken off the teeth and badly, and sometimes completely wrongly, stuck back. The removal and replacement of these can take as long as five hours of nerve-racking work per cusp, so it is clear that in the interests of economy and the preparator's sanity, fractures of this nature should be avoided.

When all cleaning and strengthening of the jaw is completed, the solid contents of the acid, washing water and that on the grid are examined to make sure that there are no loose teeth or pieces of bone or tooth among them. After this they may be thrown away. The specimen, on its grid, is returned to acid and the treatment outlined above repeated until the jaw is freed from the block, then, the whole is placed in a vessel of water. The block, while still under water, is raised slightly above the grid and inverted; the jaw will fall off onto the grid. It is washed thoroughly in this condition until all salts are removed. The residual block of matrix should be returned to acid and allowed to dissolve completely in order to make sure that it contains nothing else of interest.

The specimen is dried and final preparation can then begin. It is now obviously too small to be safely held in the hand for preparation. It should be lowered, with the side to be treated uppermost, into a small pool of molten polyethylene glycol 4000 on a piece of hardboard of suitable size. A rectangle 8 cm by 2.5 cm is about right for most specimens of this type. The wax should be on the shiny side of the hardboard and the specimen is most safely handled with vacuum tweezers, the nozzle of which has been shod with a piece of polythene capillary tube (see Chapter 2). Where no vacuum tweezers are available, a pair of light swan-necked steel forceps, the ends of which have been shod with the same kind of tube, may be used.

Held half submerged in the polyethylene glycol wax the specimen is easy to clean. The wax is not greatly affected by methyl ethyl ketone or ethyl acetate and far more force can be applied with needles and brushes than it is safe to use on an unsupported specimen. It is possible using this method to clean completely very small teeth. When one side of the specimen is finished the jaw is freed from the wax by sinking the hardboard in warm water or by the use of the heated tool described in Appendix I.

The water method is safer if the jaw is very fragile. After drying, the specimen can be mounted the other way up and that side treated.

Depending on the physical condition of the jaw and teeth the polybutylmethacrylate may be left on or removed and final treatment with polyvinyl acetate emulsion carried out.

Impregnation in this case is best done by brush under a microscope as no excess can be tolerated. If the polybutylmethacrylate must be left, the high gloss can sometimes be reduced by carefully painting the surface with emulsion.

Group 4: Isolated but Articulated Joints: Ankle, Wrist, Knee, etc.

The disarticulation of joints, both small and large, is possible by the use of acetic acid, provided that the matrix will either soften or dissolve in it.

In this type of work a cast of the specimen before preparation is essential in order that, by suitable markings of the elements on both the cast and the specimen, there is no confusion as to which is which when separation is achieved. The cast must show all sides of the fossil, for very often when the bones are disarticulated, what appeared to be two or even three separate entities in the original state, turn out to be parts of the same bone. The cast should also be coloured to distinguish between bone and matrix. The marking on the specimen may be done with either black or white drawing ink; depending on the colour of the bones. After each period of wet treatment all the markings should be examined, and if there is any sign that they are becoming indistinct they should be renewed. The type of marking is a matter of choice. Numbers or letters or a combination of both may be used or when a bone can be identified with any certainty, its anatomical name may be written on it. No piece of bone, no matter how insignificant, should be left unaccounted for. Those too small to be marked may be accounted for by making an enlarged sketch of the surrounding area marking the position of the piece and writing an identification number on this. When the fragment in question is attached to no other bone it will eventually be freed from the rock and may then be placed in a glass tube bearing its number. Should it turn out to be part of a bone which, in the beginning, was covered by rock this fact must be noted. It is very far from a waste of time to make and keep notes on the progress of preparations of this kind.

The exact method will depend on the size and condition of the specimen. When large joints such as are made up of radius, ulna and humerus or tibia, fibula and femur of anything but very small animals are involved, it is usually enough to coat the exposed bones and support the fossil in a suitable tray. The course of preparation follows the pattern outlined above. With the more complex joints, such as the carpus and tarsus it is as a rule better to limit the acid attack to one side only by using the rubber coat and fibreglass resin technique. Owing to crowding of the bones it may not be possible to achieve a safe separation from one side and in such a case, at a suitable time, the protective coating may be stripped off and reapplied to the opposite face of the joint. By this means there is a controlled solution of the rock in two directions and the easy release of the bones from each other can be effected.

As in every type of chemical preparation a certain amount of the undissolved rock has to be removed mechanically. The important thing to bear in mind is the size of the tool used. It should be narrow enough in all dimensions, to avoid the possibility of applying force to the outer edges of the bones while excavation is going on between them. Force, other than light scraping and pushing, should not be used if the matrix in the area is soluble in the acid. It may be very slowly soluble so patience is required, because the application of enough force to break up a hard matrix in a narrow cavity is always attended by the possibility that the tool will slip and thereby cause damage. The whole object of chemical preparation is to reduce the amount of violence to which the specimen is subjected, especially in areas where the bones are close together. The most difficult mineral to deal with in these circumstances is iron pyrites. Fortunately it does not often appear between bone surfaces as a solid sheet, and gentle work with a needle and repeated immersion in the acid will usually dispose of it, but it does take time. If the pyrites is absolutely solid, then the bones which it is holding together cannot be separated chemically. The only hope in such a case is when there is enough room between the bones to allow the use of a diamond dental drill to break up the obstruction. An air turbine type of handpiece is of great use in such circumstances for, as has been said before, it will stall before the drill can slip.

The complete removal of the matrix from slabs containing tarsal and carpal bone and even the ends of the associated long bones has the very obvious advantage that the joint may be reassembled without distortion; further in the case of young mammal bones, very often the epiphyses become detached after death and are separated from their shafts and may be below them. Mechanical development in which the bones are only partly revealed frequently results in their being overlooked. By chemical means, in a suitable matrix, they can be extracted without damage, and replaced on their correct shafts, as the sutures will give a positive 'fit' between the two parts of the bone.

Group 5: Complete and Partial Skeletons

Complete and partial skeletons of vertebrates may be collected as single bones or as associated bones in blocks or in nodules. When the bones are separated the method of treatment is, of course, the same as for other isolated bones, but any field data coming with them should be transferred to the specimen, using the appropriate ink.

Specimens in blocks of matrix or in nodules are simpler to deal with if the rubber coating and fibreglass resin method is used. A traditional preparation exposing only part of the bones and leaving them partially embedded in rock can be achieved by stopping the acid preparation at the required point and after thorough washing to remove unwanted salts and excess acid, either cutting back the fibreglass resin coat to give a neat finish, or removing both resin and rubber. The specimen may then be backed with a new application of chopped strand glass mat. If it is thought that at any time it may be necessary to remove the backing, the glass mat may be impregnated with a solution of a soluble resin such as one of the 'Butvars' in place of a polyester. Whenever this is done the fact should be written in some form of indelible ink on the back of the specimen, for if at some future time further acid development is required this type of coating is not suitable. Backing is not always necessary for safe display or storage; it depends on the condition of the block. For the complete removal of the bones from the matrix the procedure is as outlined for joints. Newly exposed bone is treated with polybutylmethacrylate coating and fibreglass resin casing is cut away to make access to the bones easier. When a bone is freed it is very thoroughly washed, the external coating of polybutylmethacrylate is removed and the specimen is impregnated with a 1:3 mixture of polyvinyl acetate emulsion and de-ionised water.

It occasionally happens that when a nodule containing fossil vertebrate remains is split open, the bones break through their middles and what is seen is a section through the vertebrae and limb bones; it is possible to extract the sections of the bones and to reassemble them, as was shown by the development of a partial skeleton of a small *Scelidosaurus*(?) which was in this condition (Rixon, 1968). A synopsis of the method used is as follows.

When the specimen was received the nodule was in two halves. A longitudinal section through one femur and part of the other could be seen. Adjacent to these and the vertebral column were sections through bones which, from their positions, were fairly obviously parts of the ilia. The vertebrae had been split so that the centra and parts of the zygapophyses were in one half of the nodule and the neural spines and other parts of the zygapophyses were in the other. On one side, a series of ribs had been broken so that part was in one side of the nodule and the counterpart in the other. Down one side were impressions of some large skin flaps or scutes and part and counterpart were on opposite halves of the nodule. On one side these impressions had small pieces of bone in their centres. In the area of the ribs was a crystalline deposit which appeared to contain vegetable matter. On the split faces of each half of the nodule there were ammonites and impressions of ammonites (Plates 9–11).

The specimen was first examined by an authority on ammonites so that these very important zone indicators could be identified. Samples of the supposed vegetable matter were passed to a palaeobotanist, but proved to be unidentifiable.

Both split faces of the nodule were cast. This was done primarily to record the ammonities. The bones were in such low relief that as far as they were concerned, the casts were useless. Photographs of both halves of the specimen were taken with a scale included and these were then enlarged to natural size. The two halves of the specimen were carefully examined to determine

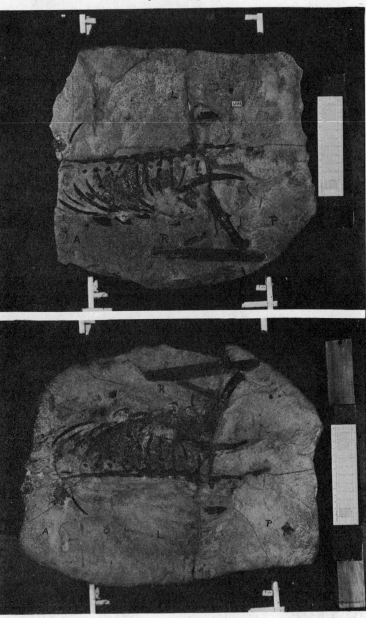

Plate 9. *Scelidosaurus*(?). Two halves of the nodule A and B showing the distribution of the pieces of split bone contained within each block. (By permission of The Trustees of The British Museum (Natural History).)

Development by acetic acid

Plate 10. *Scelidosaurus*(?). Half nodule B, after the bone content of A had been extracted by dilute acetic acid and then cemented to the counterparts in this block. (By permision of The Trustees of The British Museum (Natural History).)

which piece on one block fitted to which on the other. Each separate part of the same bone was marked with the same code number on the two faces of the specimen and these numbers were then written on the appropriate photographs.

Further examination was carried out to discover in which half of the nodule the neural arches were embedded. All visible bone and the sides and bottoms of the half nodules were treated with polybutylmethacrylate solution. When this had dried, the bone-bearing faces were covered with strong tissue paper held in place with a water-soluble glue at the edges. They were then inverted in a tray of sand, and the sides and bottoms given a very light coating of silicone fluid. Each half was then given two coats of rubber latex which was applied to all surfaces except those resting in the sand. It was found that, owing to the water repellent properties of the silicone, the first coat was not easy to apply and it is possible

108 *Chemical Methods of Development*

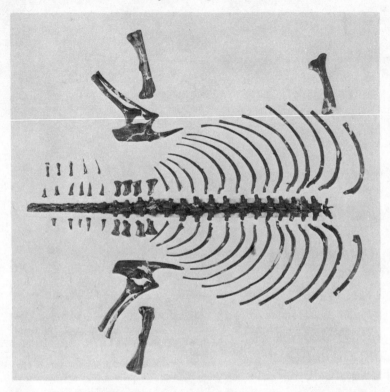

Plate 11. *Scelidosaurus*(?). 'Exploded' view of the bones recovered by further preparation of nodule B. (By permission of The Trustees of The British Museum (Natural History).)

that polyvinyl alcohol would have been a less troublesome separator. When the rubber had hardened, a dough of jute floc and rubber latex, as previously described was built up on each piece to the thickness of about 6 mm.

After this had set, a fibreglass resin basket was built over it. Twenty four hours later the two halves of the nodule were turned over, the rough edges of the fibreglass trimmed off, and the tissue paper removed.

The skin or scute impressions which contained no bone were on that half of the nodule which contained the neural arches. These impressions were surrounded by a flexible dental wax and into

Development by acetic acid 109

this dam a clear polyester resin was poured. When it had set, saw cuts were made in it with a diamond dental wheel so that each impression was on a separate block of plastic, the wax having been previously removed.

This half nodule was then placed in 10 per cent (v/v) acetic acid, and the familiar cycle of solution, washing, cleaning and strengthening carried out until the bones were freed. As soon as each piece could be taken off the block it was stuck onto the appropriate section of bone on the other half of the nodule, using a cement made by dissolving perspex (polymethylmethacrylate) in ethyl acetate. In time, all of the bone from the neural arch-containing half was transferred to the other half and the scute impressions came away on their blocks of plastic.

It was decided that it would be better to remove the scute impressions on the remaining block mechanically. This was done by making a cut with a dental diamond saw at about 6 mm away from the specimen on both sides of the scute rows. Similar cuts were made between the scutes. The outside surface of the nodule was chiselled away until a striking platform was formed which had the same depth as the saw cuts between the scutes. Each scute was struck off separately by placing a broad chisel against the bottom of the striking platform and hitting the chisel sharply with a hammer. This type of blow is known to stonemasons as a 'stunning' blow for the object is to split the stone in the direction of the hammer blow and not to cut it. The cuts stop the shock waves from breaking the stone in other places.

This half of the nodule was then submitted to acid treatment. As this proceeded, bones were removed as they came loose and new bone coded. At intervals new photographs were taken to record new finds. The final treatment of the specimen was done with polyvinyl acetate emulsion.

The result of this operation was that both femora and the entire pelvic girdle were recovered as well as the sacral ribs. Several vertebrae, not at first visible, as well as those that could be seen were extracted together with the one scapula (Plate 11). It was established that the animal had three types of skin ossicles and that ossified tendons were present in the back.

The distortion in one femur and several other bones was corrected by taking them to pieces through natural fractures and reassembling them. By these means what was only an interesting specimen was converted into one of scientific importance.

What has been said above should have indicated that one of the advantages of acid preparation is the recovery of parts of the skeleton which would not be found by mechanical development because they were situated below other bones and there was no immediate sign of their existence. In mechanical development even if their presence were revealed they could not be fully exposed without the destruction of all or part of the overlying bone.

It is also not uncommon for bones to become displaced after the death of an animal and before fossilisation occurs. The basal segment of the atlas vertebra and a perfect stapes have been found in the pterygoid region of a dinosaur skull. Earlier preparators, although showing great skill in the hammer and chisel development of the specimen, had destroyed part of the atlas segment, probably because they thought that it was a fragment of no significance or did not recognise it as bone. It must be said that this bone was so like the matrix in colour that the coloured pre-acid preparation cast does not record the presence of bone at this spot.

Group 6: Cave Breccias, Fissure Fillings, Bone Beds

Not the least important role of acetic acid is its use in extracting the sometimes tightly packed small vertebrate remains from these types of matrices. It has resulted in making some specimens, which were rare fifty years ago, now fairly common. Prior to its advent the only technique available was of the 'smash and hope' type, in which the rocks were mechanically broken up and laboriously sorted for specimens.

Cave breccias should be approached with caution, for it is in cave conditions only that the phosphates in bone commonly become replaced by carbonates and such bones will not survive immersion in acid. A small piece of breccia should be placed in the usual acid solution and allowed to disintegrate. The bone content will be almost completely destroyed if there has been any heavy replacement. In any case when bone does survive it should be examined microscopically for any signs of erosion.

Provided that the bones contained in these types of matrices are in a normal condition their extraction can be undertaken in several ways.

(*a*) Blocks may be placed in acid and allowed to break down. The insoluble residue is washed free of acid and salts and passed through a series of sieves. There will be some loss of specimens.

(*b*) The matrix may be supported on sieves made of a material which is not attacked by the acid and allowed to dissolve on them. Old terylene curtains stretched on wooden frames glued and treated with epoxy resin do very well, or plastic mesh can be used. Fewer specimens are lost in this way.

(*c*) The blocks of matrix may be enclosed on all but one side in rubber and fibreglass resin and the specimens picked out as they become loose. By occasional washing and drying and the use of polybutylmethacrylate solution, it is possible to recover all specimens.

Separation of the specimens from other insoluble material can be done with heavy liquids (Lees, 1964), but this is expensive and all such liquids are more or less toxic (see Chapter 6).

Formic Acid

As shown in a paper by the present author (1949) formic acid may be used in place of acetic acid for the development of fossil vertebrates. The use of this acid requires care and it should not be used in concentrations above 10 per cent (v/v). It will dissolve some matrices which acetic acid either will not dissolve or is slow to dissolve. A mixture of formic and acetic acids will destroy bone, so if a change is made from one to the other in course of preparation, the first acid must be thoroughly washed out before the second is used. The physiological hazards are higher with formic acid than with acetic. It can cause bad skin burns and even in the diluted form should not be handled without rubber gloves. It should never be used outside a fume cupboard as the vapour can cause eye damage.

The two acids discussed above are those which have been found to give the most consistent results. It may be said that provided that they are regarded as tools and not magic spells they will always produce a good result on suitable material. Such failures as have come to notice are due, as a rule, to one of the following causes.

(*a*) Leaving the specimen in the acid too long without strengthening newly exposed bone.

(*b*) Using an unsuitable coating medium, or overdoing the coating.

(*c*) Failure to wash out all the salts after preparation.

It cannot be over-emphasised that the acids are as much tools, in this context, as are hammers and chisels and the user must learn to control them.

Thioglycollic Acid

Howie (1974) describes the use of a solution of 5 per cent thioglycollic acid and 0.9 per cent calcium orthophosphate in de-ionised water for the removal of a haematitic matrix from the cranium of a fossil fish. Thioglycollic acid converts the ferric ions in the haematite to the ferrous state and this, in turn, is converted to a soluble ferrous salt. The addition of calcium orthophosphate was made to obviate the possibility of the removal of phosphates from the bone. After treatment in the above solution, the specimen was immersed in a 5 per cent solution of ammonium hydroxide and then washed. A layer of the matrix had softened and was removable by brushing and an Airbrasive charged with powdered sodium bicarbonate. Howie found that the best material for the consolidation of the bone, during and after preparation, was a solution of polystyrene in ethyl acetate. By the repeated cycle of acid treatment, washing and removal of softened matrix he was able to expose all the bone.

This is, without doubt, the most important advance in the preparation of vertebrate fossils made in the last twenty years. As Howie notes the extension of its use to all forms of intractable iron bound matrices is almost certain to succeed. Congratulations are due to this young worker and it is to be hoped that he will apply his very considerable talents to the solution of other problems in the field of chemical development.

Other Acids Used to Develop Vertebrate Fossils

Techniques based on the use of other acids for the development of vertebrates have been tried, but do not appear to have been used extensively. They have been evolved to deal with special cases.

Citric Acid

A solution of this acid in water will attack calcium carbonate. Carbon dioxide is evolved and calcium citrate is is formed. Calcium citrate is insoluble in water and the theory is that there is an initial attack on the surface of the bone but this is stopped almost at once by the formation of the insoluble citrate. This is true, but the same salt is also deposited on and in the matrix and all action stops very quickly. Calcium citrate is not very easy to remove so this technique is not very efficient.

R. Croucher of the British Museum (Natural History) has a variation on this method in which glycerine is included in the solution to slow up the deposition of citrates. The details of the technique have not so far been published, but it can be said that he has had success with it on material which would not stand up to acetic acid treatment. My thanks are due to him for his permission to mention his unpublished technique.

Hydrofluoric Acid

It has been suggested that this acid could be used to develop bones. I have not found it to be a good method, as bone so treated becomes very brittle. It is possible that the fault does not lie in the method, but in the material on which it was used, an assortment of scrap bone fragments. This acid will dissolve glass so must not be used in vessels made of it.

Mixed Hydrofluoric and Hydrochloric Acids

G. McGeevy and Prof. Rochow of Harvard University developed a mixture of N hydrochloric acid and dilute hydrofluoric acid to remove iron oxides and silicified matrices from fossil bone. I have no experience of their technique, but good results are claimed for it.

Hydrochloric Acid

This cannot be used to extract bone preserved in a normal way from a matrix, but can be used to destroy bone and leave a natural mould in silicified matrices. This method has been utilised in the study of ostracoderms in which the ornamented side is covered by such a matrix. A cast of the fossil in its original state is made and the matrix tested by long immersion in hydrochloric acid to make sure that it will not be weakened by the treatment.

If the matrix passes the test the specimen is placed in hydrochloric acid and the bone is allowed to dissolve. For reasons of safety, a 25 per cent (v/v) solution in water should first be tried and if this is not strong enough the concentration is increased to a point at which the bone will dissolve. Any undissolved bone at the end of the treatment will be in a weakened and finely divided form and can, as a rule, be washed out of the ornament with a gentle stream of water. Sometimes it may be necessary to use brushes and fine needles. Specimens so treated must be very well washed to remove all excess acid after treatment.

The Elgin Sandstones contain a wealth of vertebrate material, but with very few exceptions the bone is so rotten that it cannot be developed mechanically. The rock, is however, very tough and will stand immersion in hydrochloric acid. Dr A. Walker of the University of Newcastle-upon-Tyne has exploited these facts to the full and has produced a large number of casts of very elaborate skulls and bones by destroying the rotten bone with hydrochloric acid and using the natural moulds so formed.

The casts are made from polyvinyl chloride paste which takes an excellent impression, is flexible and very durable. This substance is further discussed in the chapter on casting.

Chemical Methods Applied to Invertebrate Fossils

Hydrochloric Acid

In certain localities the chemical composition of the fossils is changed to silica while the matrix still contains enough carbonate for it to be broken down by acids. With the exception of hydrofluoric, almost any acid which does not form an insoluble calcium salt could be used. Hydrochloric acid is the most convenient and is generally the one chosen. It should be diluted to the lowest concentration which will effectively dissolve the matrix. To use the concentrated acid is both dangerous and wasteful and, except for some palaeobotanical techniques, the commercial grade is good enough. Owing to the highly corrosive nature of the concentrated acid it should be treated with care when handled in bulk. Rubber gloves, aprons and a face shield should be worn.

When the diluted acid is used cold, as it normally is, polythene bowls or tanks are suitable containers in which to carry out the solution of the rock.

Chitin is not destroyed by either hot or cold dilute hydrochloric acid. The most remarkable specimens in chitin known to me are those developed with the hot dilute acid from some small ironstone nodules by Prof L. J. Wills (1959, 1960). They are the remains of scorpions and show minute detail. His papers also give an account of his use of polyester resins to hold the fossil and methods for recording the exact position of each fragment.

Palaeobotanists and others use dilute hydrochloric acid to extract specimens in the form of carbon from limestones. This element is also not attacked by dilute hydrofluoric acid and this is used to free such specimens from silicious rocks which will not dissolve in hydrochloric. Vertebrate remains are sometimes preserved as carbon.

As with all chemical preparations the specimens must be washed completely free of acid and salts before drying and sorting. Separation of the fossils from the dissolved matrix may be effected by any of the methods already suggested for the recovery of small vertebrates after acid development or those described in Chapter 6.

Acetic Acid

Phosphates will not dissolve in dilute acetic acid. When a fossil has this composition either normally or due to replacement, this chemical can be used to dissolve it out of limestone. Papers by various authors describing its use may be found in the literature dealing with brachiopods and conodonts.

Hydrofluoric Acid

Besides the uses already mentioned this acid, in rare instances, is used to extract foraminifera from rocks. It should be used diluted and only as a last resort.

Chemical Methods of Development

There are some chemicals which are used to help remove matrices from fossils by physical action and some in which there is a combination of this and chemical action. The most common are listed below.

Sodium Hexametaphosphate

A strong solution in water will loosen certain hard clays, sands which are cemented with clay and some which are loosely cemented with iron compounds. Specimens from Abbey Wood are an example of the last case. They are completely cleaned by this solution and the iron-stained sand turns white if the immersion is long enough. Dinosaur bones from a Wealdon Clay pit have been treated in this way and some cave earths can be removed from specimens with this chemical. The cruder forms seem to be more effective than the purer forms hence the commercial flakes are better for this purpose than the crystalline powders. When making a solution of the flakes it should not be stirred by hand as the partly dissolved material becomes as sharp as broken glass. A solution strong enough to be effective feels slippery if rubbed between the fingers. Although no damage has been suffered by any specimen treated with this substance, tests should be done on unimportant fragments before using it on a whole collection. It forms a complex with calcium ions and could damage some invertebrates. Specimens must be washed after preparation.

Sodium Thiosulphate

If a permeable matrix is heated in a supersaturated solution of this salt and left in it while it is allowed to cool slowly, crystallisation will occur. Large crystals will form and break up the rock. The heating and cooling cycle may have to be repeated several times. This is an old technique and used on the right type of rock can give good results. Progress can best be checked by passing the solution, while still hot, through a sieve and observing what is happening to the rock and the specimens. All specimens should be washed thoroughly after treatment.

Hydrogen Peroxide

This has been in use for a long time to break up rocks containing small or micro-fossils. It presumably works by the formation of oxygen bubbles within the pores of the rocks. The strengths used vary from 20 vol to 100 vol according to the needs of the worker. From the point of view of safety it is better to use the lower concentration whenever possible, as the storage of 100 vol hydrogen peroxide requires particular care to make sure that it cannot become overheated. The same applies to the 20 vol, but this has only a fifth of the gaseous content of the 100 vol. Both concentrations can cause skin damage. The action of hydrogen peroxide can be increased if a small amount of ammonium hydroxide is added. The ammonia solution should be diluted and added drop by drop from a pipette, as the action can be very violent and heat is produced.

Sodium Hypochlorite

Hoffmeister (1959) suggested the use of a solution of sodium hypochlorite for the extraction of specimens from coals and carbonaceous materials. He named a proprietary solution, 'Chlorox', which he said contained 5.25 per cent of the hypochlorite. Other good quality household bleaches have been found to be effective. His technique replaces a previous one in which Schulz solution was used. Although the bleaches may be used on almost any fossil, Schulz solution would damage almost anything but plant remains. The bleaches are used cold and once more the broken down matrix and its contents need to be thoroughly washed, because a large number of bleaches contain sodium chloride as well as the hypochlorite. Sodium hypochlorite will break down matrices other than coals, and is worth trying on any rock which proves to be difficult. As always, preliminary tests on small samples are advisable. Contact between acids and bleaches must be avoided as chlorine is evolved which even at low concentration is dangerous.

Pyridine

This liquid has been used to break up cannel coals. It has a highly offensive smell which is difficult to remove from specimens which have been immersed in it. It resembles benzene in some ways and should not be used except when nothing else will do. Cannal coals and some oil shales will disintegrate in chloroform, which although it is physiologically harmful, is highly volatile and is less offensive in every way than pyridine.

Disintegration of Matrices by Boiling

Some matrices, especially those derived from coagulated clays will break down if they are boiled in water. The addition of small amounts of certain chemicals is advocated by some workers; washing soda (sodium carbonate) or caustic soda (sodium hydroxide) are most commonly suggested. If anything but plain water is used a test should be carried out on a small sample, to make sure that the fossils suffer no damage, before the treatment is applied to the main batch. After such treatment the residue should be throughly washed first in tap water and then in distilled or de-ionised water. Washing soda and caustic soda will attack aluminium, so that vessels used for boiling should not be made of this metal.

When the matrix has been broken down and washed it is either passed through a series of sieves of different mesh size or spread out thinly in large porcelain or enamelled trays to dry, before separation of the unwanted mineral component from the fossils is carried out by flotation in heavy liquids.

Specimens preserved as Calcium Carbonate in Chalk or loosely cemented Limestone

The preparation of this type of fossil is usually done by mechanical means and it has already been said that the S.S. White Airbrasive makes the cleaning of chalk-covered echinoderms very easy. In the case of small highly-ornamented fossils, a chemical method has obvious advantages, since the risk of breakage due to handling is reduced. Unfortunately there is none which, in the writer's experience, will give good results in every case.

The ornament of polyzoa and echinoderms in a chalky matrix have been cleaned out successfully by a saturated solution of salicylic acid in water. Owing to its low solubility in this solvent (1:500) there is little fear of the acid damaging the specimen and long immersion is usually safe, however, the specimens should be washed after treatment. Papers have been published describing the use of solutions of oxalic and citric acids to develop fossils of this nature.

The whole question of the development of calcareous fossils

from calcareous matrices is wide open for research. It will require some medium which will act selectively upon carbonates which differ from each other only slightly. The problem is difficult, but there is increasing hope of its solution as more and more young workers trained in chemistry are becoming interested in the development of fossils. If it can be made possible it will be a major advance.

The Use of Chemicals in Special Cases

Occasionally specimens are found in which the bulk of the matrix is of the same chemical composition as the fossil, but between the matrix and the fossil there is a thin layer of material which is of a different form. This can happen when fossil grubs, or other stages of an insect's life cycle, are found as silica contained in pieces of fossil wood which are in the same condition. Between the insect remains and the wood there may be a layer of ironstone. This can be dissolved by boiling it in dilute hydrochloric acid and a space is left between the insect fossil and the wood. The insect remains will now either fall out or they may be developed by removing the fossil wood by mechanical means. The animal fossil is in no danger as there is a space between it and the petrified wood, which will prevent any violence which is applied to the wood being transmitted to it.

Preservation of Lithological Evidence

It is always likely that at some time in the future a geologist may become interested in the matrix in which a fossil has been found. The removal of a sample of matrix before chemical development will take care of this eventuality. All such samples should be fully labelled and stored with the specimen to which they belong.

The Storage of Chemically-developed Fossils

The effects of chemical development upon the mechanical strength of fossils varies with the type of specimen. In some cases in which there is a considerable quantity of compact bone the specimen, by consolidation, both during and after treatment may be made less subject to future damage than it would have been had it been developed mechanically.

Chemical preparation is, however, mainly used to develop specimens or parts of them to a degree of completeness which is not possible by mechanical methods. Before the use of acetic acid, it was never possible to study all aspects of every bone of an ichthyosaur skull in any one specimen, except by serial sectioning (Sollas, 1916). Now such skulls can be taken completely to pieces, and if it is done properly, even such delicate areas as the sutural contacts between the palatines and pterygoids can be preserved. While many bones from the skull of this reptile are robust, by any standards some are not, so require very careful packing for storage, and need gentle handling when being studied.

Cotton wool (cotton) should not be allowed to come in contact with delicate specimens. It may however be used as an underlayer. Paper tissues can be used with safety in almost every case.

To prevent movement of a specimen in its storage box it is sometimes necessary to place it in a shaped hollow well within the container. Expanded polystyrene can be bought in a wide range of thicknesses and this can be carved easily with either sharp or hot tools into any form; it is therefore a suitable substance for this purpose, provided that the specimens are absolutely free of solvents and have not been treated with anything which will stick to the expanded plastic. Should a mistake be made and the specimen does stick, it may be freed by carefully cutting the plastic, as close to the specimen as possible, and lifting this section plus the fossil out of the box. The polystyrene which adheres to the specimen may be removed first, by the careful use of a hot wire and finally by dissolving the residue in ethyl acetate applied with a brush. Methods for shaping expanded polystyrene are described in Appendix I.

There are many forms of synthetic rubber foam sheetings on sale and most of them can be used to construct protective packings for fragile specimens, but before a type is put into use it should be checked for undesirable properties such as high acidity or a tendency to stick to an object that has been resting on it for some time. The latter eventuality can be guarded against by putting tissue paper between the fossil and the packing.

It is not possible to give more than a general outline of procedure for packing for storage of this type of specimen, as each one requires personal attention. The important thing is to keep the specimen immobile in its drawer, if such is its place in the store.

Very small specimens, such as small mammal or reptile teeth and jaws, require special attention. Teeth are often so small that they cannot be stored in tubes alone because they might get broken if no packing is used or get lost in the packing if it is used. They can be stuck, with a water-soluble glue to the bottom of cardboard cavity slides fitted with a sliding glass top, but almost certainly the first scientist who wishes to study them will not be able to see all that he wants and they will have to be removed and replaced. The more often this is done, the more likely it is that damage will occur. Such specimens have been stuck to the ends of bristles which in turn, were stuck into the corks of glass tubes. This is good enough up to a point, but there is not much contact area for the adhesive, bristles are very springy, and it is not impossible that an unfortunate contact with the bristle will catapult the specimen away, and somebody will have to spend hours looking for it. The ultimate answer to this problem has not yet been found, but the method detailed in Appendix I obviates at least part of the difficulty.

To avoid unnecessary handling of small jaws and similar specimens, they may be mounted on plastic frames constructed so that they fit into small rectangular plastic boxes. Suitable plastics are Perspex or cellulose acetate sheet which is 1.5 mm or less thick. Both of these materials are easily made malleable if they are gently heated. Details for making this kind of mount will be found in Appendix I.

References

Camp, C. L. and Hanna, G. D. (1937) *Methods in palaeontology*, Berkeley: Univ. of California Press.

Hoffmeister, W. S. (1959) ' Sodium hypochlorite, a new oxidising agent for the preparation of micro-fossils', *Oklahoma Geol. Notes*, **20,** 34.

Howie, F. M. P. (1974) 'Introduction of thioglycollic acid in preparation of vertebrate fossils', *Curator*, **17,** 159–65.

Lees, P. M. (1964) 'A flotation method of obtaining mammal teeth from Mesozoic bone beds', *Curator*, **7,** 300–6.

Macadie, C. I. (1966) 'Use of flexible fibre light guides in preparative palaeontology', *Mus. Journal*, **66,** 215–16.

Rassetti, F. (1941) 'Action de l'acide oxalique sur les calcaris fossiliferes' (abstract), *Assoc. Canadienne-Francaise. Adv. Sci. Annals*, **7,** 91.

Rixon, A. E. (1949) 'The use of acetic and formic acids in the preparation of fossil vertebrates', *Mus. Journal*, **49,** 116–17.

Rixon, A. E. (1965) 'The use of new materials as temporary supports in the development and examination of fossils', *Mus. Journal*, **65,** 54–8.

Rixon, A. E. (1968) 'The development of a small *Scelidosaurus* from a lias nodule', *Mus. Journal*, **69,** 315–21.

Sollas, W. J. (1916) 'The skull of Ichthyosaurus studied in serial section', *Phil. Trans. Roy. Soc. (B)*, **208,** 63–126.

Steen, M. C. (1931) 'The British Museum collection of amphibia from the middle coal measures of Linton, Ohio', *Proc. Zoo. Soc. London*, **1930,** 849–91.

Toombs, H. A. (1948) 'The use of acetic acid in the development of vertebrate fossils', *Mus. Journal*, **48,** 54–5.

Toombs, H. A. and Rixon, A. E. (1950) 'The use of plastics in the transfer method of preparing fossils', *Mus. Journal*, **50,** 105–7. (1959) 'The use of acids in the preparation of vertebrate fossils', *Curator*, **2,** 304–12.

Wills, L. J. (1959) 'The external anatomy of some carboniferous scorpions. Part I, *Palaeontology*, **1,** 261–81, pls. 49–50. (1960) 'The external anatomy of some carboniferous scorpions. Part II', *Palaeontology*, **3,** 276–332.

6
The Concentration of Small Fossils from Bulk Matrix

A little has already been said on this subject in Chapter 5 but, because it is one of the commonest techniques employed in the extraction of small fossils from matrices in which they occur in abundance, the methods and apparatus used are worthy of further description.

The first essential is that the matrix containing the specimens should be reduced to a finely divided state so that it may be separated from the fossils either by sieving or by flotation. This may be done either by the chemical methods already described or mechanically.

Apparatus for the Pulverisation of Rock Samples

The simplest apparatus for reducing a matrix to a finely divided state is the pestle and mortar. The types used in chemical laboratories have enough strength to withstand the force required to pound the softer limestones, but the harder rocks can only be dealt with in an iron mortar, using an iron pestle. The method is to break the sample into small pieces with a hammer and then place a few of these at a time into the mortar and crush them until the rock is reduced to the required grain size.

More refined apparatus has been devised in which the fragments of rock are placed in a container with a heavy steel roller and the container rotated mechanically; but the installation of this is only worthwhile in a department dealing with a continuous series of samples which have to be broken down.

The mechanical breakdown of rock is only used when there is no other method of achieving the required result and is confined to the study of microfossils.

Sieves and Sieving

The sieve is the most common tool used to separate specimens from a granular matrix, whether this be natural or arrived at by any of the means previously mentioned. The types vary from the ordinary garden sieve, with holes about 3 mm to 6 mm square, to very fine gauges with several hundred holes per square centimetre. The coarser types are commonly used in the field for separating fossils, or fragments of them, from non-coagulated matrices such as gravels and sands; normally when the material is dry.

McKenna (1962) gives a long description of separating fossils from their matrix in the field by washing and screening them in natural running waters such as rivers and streams. His was a massive operation in which hundreds of kilograms of material were dealt with and he described how to make suitably-sized sieves from timber and cheap wire mesh. It would be as well, before attempting such an operation, to make sure that one was not committing an offence or offending anybody who has rights such as fishing or watering cattle, at points below the washing site.

Sieves for use in the laboratory may be bought from the larger suppliers of chemical apparatus, firms specialising in the supply of apparatus to geologists and mineralogists, smaller concerns who sell to naturalists or directly from the makers. The best types are those in which the base of one will fit into the top of another, so that a stack of varying mesh sizes can be built up. This kind is provided with a container which fits below the lowest sieve and a lid which fits the top one. Diameters and mesh sizes vary, for general purposes a diameter of approximately 30 cm is usual. Mesh sizes are graded according to holes per inch (2.54 cm) and a wide selection is needed. Which mesh size to buy is governed by the size of the fossils most commonly dealt with, the amount of storage space available and how much you can afford to spend. If the last two factors are of no significance, it is better to have as many sizes down to about 200 mesh as you can, for although some may only be needed once every five years, time is saved if they are immediately available when required. The mesh of these sieves is usually made of brass wire; this is to be preferred to iron wire, since most laboratory sieving is a wet process and iron rusts very quickly.

Besides the complete sieves, most suppliers sell lengths of the wire mesh from which they are made. This is very useful, for not only can it be used to repair a worn-out sieve, but if the edges of a small square are bent upwards this makes a practical small sieve which is much easier to clean than one with solid metal sides. This is an important factor when dealing with stratified samples which must not be contaminated by specimens from other layers. For use with the stronger acids, a mesh made of Monel metal is employed; since this is rather expensive it should be restricted to this use.

In many instances it is not necessary that a sieve with an accurate mesh size be used. As was briefly mentioned when discussing acid treatment, plastic meshes are available, which are manufactured for making fly screens for use in tropical countries and some very fine ones for filter cloths. Attached to suitable frames, these make very good sieves. Used in the same way, terylene curtain net has many applications; it is cheap enough for it to be used for one job only and then thrown away. Nylon mesh used to make stockings and some other articles of clothing may be utilised in the same way.

Dry Sieving

When the matrix is in a finely divided condition and will flow freely, the separation of the fossils may be done in the dry state. If it is known or suspected that the contained fossils differ in size, a series of sieves is selected in which the one with the largest holes will stop the passing of the largest specimens and the one with the smallest holes will stop the smallest. The sieves are stacked with the one with the largest holes at the top and the others in graduated series to the smallest holes at the bottom and the container is attached below the pile. Part of the sample is placed in the top sieve, which should not be overloaded. The lid is fitted and the assembly gently shaken. The stack is then separated into its individual sieves and the contents of each are transferred to a cardboard tray containing a label which refers to the geographical and stratigraphical locality of the sample and the mesh size of the sieve, then the stack is reassembled and more of the sample placed in the top and the process repeated until all has passed through. The contents of the bottom container should be examined under a microscope to establish that it contains no fossils.

The contents of each cardboard tray is examined in small quantities; about 4–5 ml at a time is a manageable quantity if it is transferred to a 5 cm square cardboard tray. The different types of fossil are picked out and placed in other trays or glass tubes and each container must of course, have an appropriate label.

The larger fossils may be picked up with fine swan-necked forceps, the smaller ones with the moistened tip of a fine paint brush. Water will usually suffice to dampen the brush, but when it does not, a very small amount of a water-soluble gum may be added. This should not be overdone or the sorted fossils will stick to each other as the gum dries out. It is a common practice to moisten the brush by putting it into the mouth; this is not a very hygienic habit and, in the case of fossils which have been treated with some chemicals and not sufficiently washed, it could be dangerous and should certainly never be done when fossils have been extracted with hydrofluoric acid.

Apparatus is available for automatically shaking sieves, but it is rather noisy in operation and is only worth the investment if a great deal of dry sieving is required as a regular part of laboratory routine.

When the sample is small, the matrix is finely divided and there is no special advantage in grading the fossils by mesh size of the sieve, the operation can be carried out with a single sieve whose mesh will stop the passage of the smallest specimens.

Wet Sieving

This is the method which is best for most fossils but there is one important matter to be considered before starting such an operation. This is that the excess water has to be carried away by the drainage system and will take with it a considerable amount of solid matter which may build up ultimately to form a blockage somewhere in the system. This difficulty can be overcome if the outflow of the sinks is passed into a sedimentation tank; drainage engineers should be consulted on this problem. Such a tank can be small and placed in the laboratory below the sink, or the drainage system from the laboratory may be carried into a large tank outside. Any type of sedimentation tank must be cleaned out regularly. Where a tank cannot be fitted or the amount of this type of work which is done in the laboratory does not warrant it, the outflow from the sieves should first fall into a large vessel which is placed in the sink. Its sides should be at least 5 cm less in height than the depth of the sink and it must not be so large as to cover the drainage hole. The water entering this vessel will eventually overflow into the sink, but the heavier fraction of the solid it carried will be deposited on the bottom and only very finely divided material will be carried out with the water.

When a stack of sieves is used, the bottom container and the lid are not required. The base of the stack should be stood-off from the bottom of the sink or the overflow vessel; this can be done with wooden battens but if the assembly is not heavy enough to stop them floating they will need to be weighted; strips of lead sheeting, or any other substance which is heavy and is not affected by water may be used. The top sieve is loaded with a part of the sample as described for dry sieving and water is run through the sieves from the top downwards. The stream should not be so strong as to endanger the fossils, and it should be broken up into a spray by plastic 'roses' made for use with watering cans, sprays made for shampooing hair or those intended for attachment to kitchen taps. A plastic spray is better than a metal one, because if it falls into the sieve it is less likely to damage the specimens. The spray is moved by hand over the surface of the sample until the water emerging from the bottom is clear. An effort should be made to keep as many of the specimens as possible away from the sides of the sieves, because in some types there is a depression between the sides and the mesh and it is often difficult to remove specimens from this without damage. When the washing in tap water has been completed each sieve and its contents may be rinsed in distilled or de-ionised water to remove any salts contained in the mains supply. Whether this is necessary or not depends on the dissolved content of the local water; in soft water areas it is often not worth doing.

The specimens are now left to dry on the sieves; this is best done at room temperature but heat may be used to accelerate the process if it is kept to about 50°C. The fossils are then transferred to trays and sorted as described for dry sieving.

Although it will appear to be implied that all that remains in the sieves after shaking or washing are specimens, this is obviously not the case, for any mineral matter which is the same size as the fossil will remain on the same sieve, but as a rule the specimens can be separated by hand at this stage. If a large amount of unwanted residue has been retained, separation by heavy liquids or other means may be required.

As with dry sieving, it is not always necessary to use a graded series of sieves and one that will retain the smallest specimens is enough. A spray is not then needed but the loaded sieve may be repeatedly immersed and then lifted out of a large vessel containing water.

Other Methods of Separating Fossils from broken down Matrix by Water

It is sometimes possible to extract the fossils from a finely divided matrix by methods which were used by gold prospectors. The first is 'panning' and only works well if the fossils are fairly large and the mineral fraction is finely divided, although it may be used before sieving to remove the lighter parts of any sample.

The material is placed in a bowl with sides about 5 cm high which is partly filled with water. The vessel is then rocked and swirled around, which causes the light particles to go into suspension. The water is decanted off, taking the suspended solids with it, into another larger container. The process is repeated until the agitated water is clear and no more solid can be washed out of the sample. The contents of the pan are then tipped out onto some suitable surface to dry and more of the sample treated in the same way. The vessel into which the waste has been poured should be left standing until all the solid matter has settled, then as much of the water as possible is removed, without disturbing the settled material, by syphoning or any other convenient means. The 'sludge' is now scooped out of the vessel and laid out to dry and is examined to see that it contains no fossils.

The second method is a variation on the miner's washing chute. The easiest material from which to make this is the three-sided metal covering which is used to protect electrical wiring. This is available in various lengths, and about 2–4 m is required. The holes made for the fixing screws may be filled with epoxy putty or plastic wood. A 'T' piece of plastic or metal tubing that will fit across one end is required, into the edge of the cross piece farthest from the tail a series of small holes are bored so that, if the tail is connected by rubber tubing to a tap, a series of jets of water emerge through these holes.

The conduit is then placed at an angle of about ten degrees to the horizontal and held firmly in position on wooden blocks or wedges. The 'T' piece is fixed at the higher end and the other end overhangs a sink into which a plastic bowl is placed (Fig. 8).

The sample is loaded into the high end of the chute and the water turned on and by adjusting the force of the jets and the angle of the chute, a position can be arrived at in which the specimens are separated according to size and weight down the length of the chute and most of the mineral content is washed away and collects in the bowl in the sink which will also trap any fossils which are washed over. This method has worked well with fairly large silicified molluscs extracted from limestone with dilute hydrochloric acid. Its drawbacks are that a considerable amount of bench space is required and it can take some time to arrive at the optimum water pressure and angle of the chute. If the material suggested for making the chute is not easily obtainable, then one

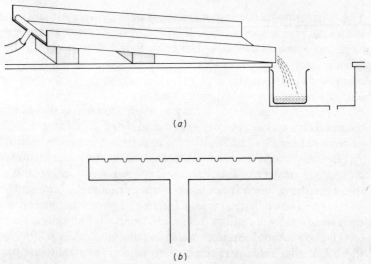

Fig. 8. (a) Washing chute for the concentration of small specimens (not to scale); (b) Detail of 'T' piece.

could revert to the methods of the prospectors and make the trough out of wood and such a structure would be improved if treated with an epoxy varnish.

Soap Solution

Howe (1941) discusses the use of soap or washing powders for the extraction of microfossils from sandy matrices. The correct amount of these materials used in warm water will cause small shells, foraminifera and ostracods to float. He does not state the exact quantity, presumably because it could differ with the type of soap or powder used. The extracted fossils were afterwards washed on a 200 mesh sieve.

A variation on this technique is practised by archaeologists who use a solution of a detergent in water, to isolate plant remains from soils. The sample is placed in this and streams of diffused air are passed from the bottom of the vessel, causing the specimens to be carried upwards in the foam and trapped in it. This method would no doubt work very well on some fossils.

Separation by Heavy Liquids

The separation of one mineral from others by heavy liquids is a technique which has been used by mineralogists for a very long time and some palaeontologists have found it to be useful but because of the high cost and toxicity of most of these liquids in my opinion their use should be restricted to samples which cannot be efficiently concentrated by other means.

Of the many liquids which may be used, bromoform is the most common. Its specific gravity may be adjusted to suit the sample under treatment by the addition of industrial methylated spirit.

The best method is to place a part of the sample into a measured quantity of industrial methylated spirit and to add to this a measured quantity of bromoform until a part of the sample floats; this is usually the mineral fraction. The floating material is then removed by decanting it off with some of the liquid. As much of the clear bromoform mixture as is possible is then poured off the solids that remain at the bottom of the vessel, the last part being poured through a filter paper supported in a filter funnel which will trap any specimen coming over with the liquid. The fossils are then washed several times in changes of the spirit and finally laid out to dry on sheets of filter paper in a fume cupboard. All the operations described should be carried out in such a cupboard and the whole of the liquids including the spirit used for washing should be preserved.

The bromoform may be recovered by mixing the spirit used for washing with the original mixture and then adding to this a large excess of distilled or de-ionised water and shaking. The spirit will dissolve in the water and, if left to stand, will float on top of the bromoform and the two liquids may be separated by the use of a separating funnel. The recovered bromoform may then be filtered and for most purposes it can be used again but in instances where there must not be the slightest chance of the bromoform being contaminated it should only be purified by distillation. This is an operation which should never be attempted by a person who has not been instructed in methods used in organic chemistry and is beyond the capacity of most palaeontological laboratories; the assistance of a properly qualified chemist working in a well equipped chemical laboratory is required. Griffith (1954) used undiluted bromoform on specimens from the Rheatic bone beds of Gloucestershire, Avon and Glamorgan. The material was first washed in water and then the matrix broken down in dilute acetic acid. The next step was to wash the whole very thoroughly in water; and then after drying it was passed through 10, 20, 30 and 40-mesh sieves. That in the 10-mesh sieve was placed in bromoform, when the mineral matter floated and was separated by decantation. The bromoform was thoroughly washed out with spirit or acetone and the fossils allowed to dry out for several days before sorting.

Carson (1953) used tetrabromoethane or acetylenetetrabromide diluted where necessary with acetone and reclaimed the tetrabromoethane from the acetone by the addition of an excess of water. The fossils concerned were foraminifera. Lees (1964) used a mixture of tetrabromoethane and *syn*-dibromoethane. Bone sank in this mixture and the mineral fraction which floated was separated by decanting.

Carbon tetrachloride may be used in some instances to separate foraminifera from a disintegrated matrix.

Let it be said once more that the vapours from all these liquids

132 Concentration from Bulk Matrices

are toxic. They should not be used outside a fume cupboard and it would be sheer folly to use them in a badly ventilated room. Any amateur who should read this book is strongly advised to leave them severely alone. Specimens which have been treated with them, even after washing with appropriate solvents should be put aside for several days, preferably in a fume cupboard, before they are sorted.

Concentration by Magnets

Mineralogists separate non-magnetic from magnetic minerals in this way and apparatus in which the finely divided sample is mechanically shaken past the poles of a magnet is available.

It is said by some palaeontologists that certain fossils may be separated from a finely divided matrix in the same way. I have never encountered a sample which responded to this treatment but this does not mean that the method is invalid, only that I have been unlucky. It is worth trying on any sample in which it is likely that the fossils contain a magnetic material or the bulk of the matrix is magnetic.

Test apparatus need not be expensive since very powerful permanent magnets can be obtained from old moving coil loudspeakers, which can often be bought at a reasonable price from shops which sell ex-radio equipment. Smaller magnets can be taken from broken moving-coil ammeters and voltmeters, but the larger circular magnets from speakers are better for this purpose.

A sample may be tested by covering the pole face of the magnet with some thin paper held in place with sticky tape. A small amount of the material under test is spread on the paper and the whole gently inverted when anything which is attracted to a magnet will remain on the paper and the rest will fall off. The retained fraction can be lifted off on the paper.

An electromagnet would be better, but they are not so easy to come by, require a D.C. supply and, like all electrical apparatus, have to be used with an eye to the potential danger of the operator receiving a shock.

A powerful magnet has many uses in a laboratory. Among others it can be used to magnetise small diameter steel rods which are useful for the recovery of small steel objects which have fallen into inaccessible places.

The Soxhlet Extractor used for Washing and Concentrating Specimens

The advisability in all cases, and the necessity in some, of removing undesirable salts from specimens has been mentioned in other parts of this book. This operation is no great problem when a few moderate-sized fossils are involved but if several dozen specimens whose maximum length is about 3 mm have to be treated there are difficulties.

If the washing is done in constant changes of solvent in evaporating dishes a great deal of bench space is required and when a volatile solvent is used this is in itself a problem.

It has been found that many fossils may be washed with safety in a Soxhlet extractor. This is a standard piece of chemical apparatus consisting of a round-bottomed flask into which is fitted the Soxhlet tube and above this is a water-cooled condenser (Fig. 9).

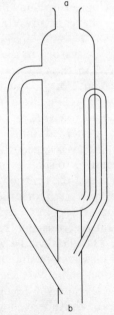

Fig. 9. Soxhlet Extractor tube. At point a, a water-cooled condenser is fitted and point b enters the neck of a round-bottomed flask.

When the liquid in the flask is heated it vapourises and the vapour passes up the wide tube and then into the condenser where it is converted back to the liquid form and flows down into the Soxhlet tube. As the liquid rises in this tube it also rises in the narrow tube which enters the bottom. When the liquid passes the bend in the narrow tube a syphon action is started and the fluid is carried back to the flask. This cycle results in there being, in the Soxhlet tube, a constantly changing supply of clean solvent and the soluble matter in the specimen is carried down to and remains dissolved in the liquid in the flask.

Standard thimbles are made into which the material to be washed is placed before it is put into the main body of the Soxhlet tube. Some are made of porous paper and some of glass with a sintered base. Both types can be used for containing fossils. The paper types give less trouble with regard to the solvent being emptied from them by the automatic action of the apparatus, but it is not easy to see what is happening to the specimen. The sintered glass type can become clogged up and the solvent is then not changing rapidly, but they have the advantage that all sides of the fossil can be observed.

Quite large fossils may be washed in these tubes. Because the condensed solvent can fall from the condenser in quite large drops, it is as well to plug the mouth of the thimble lightly with some porous material to stop them falling directly onto the specimen.

The thimbles described are too large for use with the very small fossils with which we are concerned; for these, special tubes have to be made. Some glass tubing is selected whose inside diameter is large enough for the specimens to slide in and out with ease. One end is rotated in a flame until the central hole narrows slightly. The middle of a piece of thin glass rod is heated to redness and pulled out to form two rods with pointed ends. A small area of one side of the tube is heated to redness and the red hot point of the glass rod placed onto it. The glass should weld and, if the rod is pulled, a narrow tube of glass will be drawn from the side of the original tube and if the junction between this and the glass rod is broken the end of the small tube will be opened. This action is repeated at various points on the tube at roughly 1.25 cm intervals and the tube is then cut at about 1.25 cm away from the first hole so formed (Fig. 10a).

The Soxhlet extractor

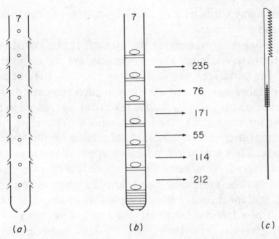

Fig. 10. Tube to contain very small specimens when washing in a Soxhlet Extracter. (a) The tube showing vents; (b) Diagram used to locate specimens during treatment; (c) Section of fretsaw blade used to remove paper plugs after treatment.

The resulting perforated tube can be of any length, provided that, when it is placed in the Soxhlet tube, its top is just below the bend in the syphon. Several tubes should be made and each one numbered with a diamond writing pencil.

When these tubes are used they are taken, one at a time and a sketch consisting of two parallel lines drawn. Between them, at the top, the number of the tube is written. A light plug of porous paper, paper handkerchiefs are a good source of supply, is pushed into the tube from the top to the bottom, or glass wool may be used in place of paper. The first specimen is then placed in the tube and covered with another plug. On the sketch the basal plug, the specimen and the upper plug are indicated and the registered number or any other identification of the specimen written level with this area. This is repeated until as many specimens as can be packed into the tube have been loaded and recorded. The top is then closed with a loose plug (Fig. 10b). As many tubes as will fit into the apparatus are filled and transferred to it, and the apparatus set to work. Provided that some safe form of heating, such as an Isomantle or a water bath is used and the water supply to the condenser is constant, the apparatus will function without attention but, as always, it is wise to check from time to time that all is well with the specimens.

When the washing has ended the apparatus is allowed to cool and the tubes containing the specimens are taken out. The procedure for unpacking the fossils is to take one tube at a time, check its number against the sketch and then remove the fossils from the top downwards. As each is taken out it is placed in a container bearing its number and this number cancelled on the sketch, so that there is no possibility of mixing up the specimens or losing one. They are then dried, by an appropriate means, while still in numbered containers from which they are eventually transferred to their numbered and labelled storage tubes.

A useful tool for the removal of the plugs from the washing tube is a section of a fretsaw blade attached to a wire, with its teeth pointing backwards (Fig. 10c). This is slipped between the glass and the plug and then turned so that the teeth bite into the paper; it is then withdrawn and the plug should come with it.

The Soxhlet extractor has many uses and has even been employed to extract specimens of mollusca fossilised in a matrix bound with a bituminous substance like asphalt.

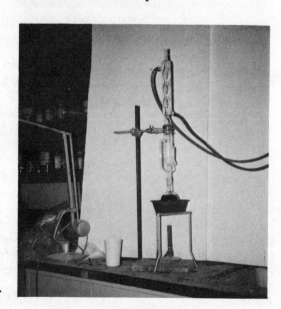

Plate 12.
Soxhlet apparatus.

In this case a plug of glass wool was placed at the bottom of the Soxhlet tube, the matrix with the enclosed specimens placed on it and another plug placed on top. Chloroform was used as the solvent; some small pieces of broken porcelain were placed in the flask to stop 'bumping' as the solvent boiled.

The specimens were perfectly white, clean and not sticky after treatment. About 300 ml of solvent was used and to have obtained a comparable result by cold washing in changes of solvent would have required very much more than this. Owing to the fact that chloroform vapour is heavy and toxic, unless a long condenser is used some may escape into the air; this solvent should be used only in a fume cupboard.

When this apparatus is in use it is necessary to take account of some potential hazards. The flask, no matter what it contains, should never be heated directly by a flame or hot plate. With water it may be placed in a sand bath. If the solvent has a low boiling point a water bath should be used and must not be allowed to boil dry. Never use an inflammable solvent if a non-inflammable one will do and on no account should ether ever be used. If the solvent is inflammable the apparatus should never be left unattended and proper fire-fighting equipment must be close at hand. It will probably help if the apparatus is placed in a metal tray large enough to contain all the solvent present, its bottom covered in an inch of sand. Toxic solvents should be used in a fume cupboard or some other efficient precautions should be taken to remove any fumes which may leak.

When salts are being washed out of specimens in this way water is used and their complete removal is indicated when the water in the flask, which should be changed at intervals, gives a negative reaction when tested for the salt in question.

If a resinous substance is being extracted the solvent in the flask should be changed before the solution becomes too viscous.

Soxhlet apparatus is made in many sizes and can be obtained with standard ground glass fittings at the ends of tubes and mouths of flasks. This type is the most generally useful as it can be used with all solvents.

References

Carson, C. M. (1953) 'Heavy liquid concentration of foraminifera', *Journ. Palaeont.*, **27,** 880–1.

Griffith, J. (1954) 'A technique for the removal of skeletal remains from bone beds', *Proc. Geol. Assoc.*, **65,** 123–4.

Howe, H. V. (1941) 'Use of soap in the preparation of samples for micropalaeontological study', *Journ. Palaeont.*, **15,** no. 6, 691.

Lees, P. M. (1964) 'A flotation method of obtaining mammal teeth from Mesozoic bone beds', *Curator*, VII, 300–6.

McKenna, M. C. (1962) 'Collecting small fossils by washing and screening', *Curator*, V, 221–35.

7
The Effects of the Decomposition of Iron Pyrites within a Specimen and Methods used for its Arrest

Some fossils, notably those from the Lias, Gault and London Clay, contain a considerable amount of iron pyrites and this can decompose. The decomposition always results in some damage to the specimen and in extreme cases its complete destruction. By tradition, specimens so affected, have been said to be 'pyritised' or 'pyritising'. This was an unfortunate choice of words because they mean the exact opposite of what actually happens. This 'disease' has long been the most serious problem in the conservation of fossils and in spite of efforts by various workers over a long period there is still no preventative treatment or cure which can be said to be absolutely effective, but there are some which, in many cases, will arrest the progress of decomposition for several years. Instances are known of fossils treated thirty years ago which still survive.

Radley (1929) published three short papers describing his work in trying to establish the cause of the breakdown. It is interesting to note that he mentions, even at that date, there was a 'microbe' theory, since recent workers tend to think that a bacillus is at least partly responsible for the trouble. Radley concluded that the disintegration was due to the action of oxygen in the presence of water converting the sulphur in the pyrites to the dioxide which, eventually by the action of water and oxygen, produced sulphuric acid and this attacked the calcium carbonate content of the fossil. Sulphates of iron were found in the products of decomposition. He also concluded that chlorides, such as salt, were accessary agents of decomposition and noted that fossils which were washed free from salt and then dried did not decompose so often or so quickly as those which were not so treated. Apart from the removal of the decomposition products from the surface of the fossil by brushing, followed by thorough washing and drying and then coating with a nitrocellulose varnish, he recommended no treatment.

Experience has shown that the removal of salt from specimens whenever possible is a good practice, but that washing a specimen which has already started to decompose will sometimes result in its complete collapse. Although cellulose nitrate varnish was probably the best available in 1929, time has shown that it ceases to be effective after a number of years.

Bannister (1933) dealt with the preservation of mineral specimens of pyrites and marcasite. He recommended that the acid produced by decomposition should be neutralised with ammonia either as a liquid in the form of the hydroxide or as the gas. This treatment was followed by thorough drying and the specimens were then coated with a solution of polyvinyl acetate in toluene. Variations on this technique are still the most widely used methods of treatment for fossils and in many cases are highly successful.

Booth and Sefton (1970) described the vapour phase inhibition of thiobacilli and ferobacilli and they suggested that storage in the presence of 4-chloro-*m*-cresol is a possible way of preserving specimens composed of or containing pyrites. Should time prove that this is the answer to the problem, a great debt will be owed to these two authors, as this treatment does not involve the use of varnishes or other coatings and the appearance of the specimen is not changed.

There are, however, practical limitations to the size of specimen to which this method, in its present form, can be applied since 4-chloro-*m*-cresol is a solid which gives off a vapour and cannot be very easily used on large specimens. The substance is corrosive hence cannot be scattered about indiscriminately. There is, however, no problem when dealing with specimens small enough to fit into tubes or small boxes. In a tube, a few crystals can be put at the bottom and covered by a plug of some porous material and the specimen placed above this. If the tube is tightly corked it should not require attention very often. When boxes are used for storage a small tube of the crystals, fitted with a porous plug, may be placed in the box with the specimen.

Since this method involves the use of a substance with which many workers in palaeontology will not be well acquainted, it is advisable that the properties of 4-chloro-*m*-cresol should be established from a suitable chemical index before it is used, thus ascertaining the precautions that will be needed with this chemical. It would be wise to label tubes or boxes containing the crystals indicating that they are corrosive to the skin and precautions are needed to avoid the introducing of small particles into the eyes.

The Symptoms of the Breakdown of Pyrites

If the breakdown has occurred near the surface of the specimen, a greyish-white powder will appear on the fossil. The appearance of this deposit is so characteristic that a worker who has seen it before will recognise it for what it is at once. There is a test which establishes beyond doubt whether pyrites has broken down; a small area moistened with a little ammonium hydroxide will, on drying, turn brick red. Occasionally the breakdown products contain yellow flecks of sulphur.

Some reptile bones, especially the limb bones of plesiosaurs and pliosaurs have a thick outer layer of compact bone in their shafts, and an inner core of cancellous material which is full of finely divided pyrites. When this breaks down there is an increase in volume of the inside of the bone and this can cause massive cracks in the outer layer and great distortion. Usually some pieces have fallen away from the fossil or can be lifted out, and the characteristic powder can be seen in the inside.

Where skeletons are preserved on slabs of rock, the breakdown may occur not only in the bones but between them and the rock and within the rock itself. Fortunately the decay of pyrites within the laminations of rocks is not common except in some types of coal. The outward signs are cracking and bulging of the surface which may break or loosen the specimen on it.

Treatment with Ammonia

Some form of airtight container is required, for small specimens a large bell jar standing on a piece of ground glass will do and larger fossils may be treated in a well-made box, preferably fitted with a glazed lid. Since ammonia is very pungent and is unpleasant to inhale, the box should be fitted with two gas outlets so that before it is opened air may be blown through one and the ammonia conducted by a tube from the other, either to the open air or via a wide-mouthed funnel into a vessel of water. If the box can be placed in a fume cupboard this precaution is not necessary, for the lid can be removed with the front of the fume cupboard pulled down as far as possible.

The treatment will be most effective if as much of the surface area of the specimen as possible is exposed to the gas. This is most easily achieved if a shelf of fine plastic mesh supported by a wooden frame is made. The shelf should not be a fixture but should be easily removable from the gas chamber; for heavy specimens it can be made of wooden slats.

The fossils to be treated are examined and if the products of decomposition can be brushed off without disturbing the surface of the specimen this should be done, but if the fossil is covered with loose pieces and a large number of interlacing fine cracks, it is better left alone, as the deposit is sometimes easier to remove after treatment with the gas. Together with identification labels, the fossils are placed on the shelf and this is put into the container. If the specimens are very fragile they should remain in their trays or boxes. A small wide-mouthed vessel is now half filled with 0.88 S.G. ammonium hydroxide and placed on the floor of the chamber, the lid is closed or the bell jar lowered, and the specimens left in the atmosphere of ammonia gas for several days; then, taking precautions against the sudden inhalation of the ammonia, the container is opened and the shelf with the specimens removed. It will be noticed that the areas in which there are breakdown deposits have turned red, and in some cases the loosened pieces are less inclined to move.

The fossils are now transferred to cardboard trays, if their condition has allowed handling without support at the first stage, and dried in an oven set at 50°C for at least one day. After drying any reddened deposit which can be removed without damage to the fossil should be gently brushed off or scraped with a needle. Any loose pieces which can be removed and replaced are stuck back into position with a quick drying adhesive. When all joints are thoroughly dry, the fossils are coated with either a polybutylmethacrylate solution or a solution of polyvinyl acetate in toluene; if available, Bedacryl 122X is to be preferred. The solutions should be thinned with toluene to a viscosity which will give good penetration into the specimen.

The coating must be done carefully and must be complete; it is advisable to repeat it several times, letting each application dry before the next is made. The safest method of application is by dropping from a brush or a polythene pipette but robust specimens may be coated by dipping. Care must be taken to avoid the specimen sticking to the bench. When the last coating has dried, the specimen may be further strengthened by filling the larger cracks with either a jute floc-based gap-filling compound or with an epoxy putty. When finished, the fossil will have a high gloss, which can be reduced if the coating is allowed to dry slowly in an atmosphere of toluene or if a small quantity of aerogel silica be added to the solution. If the latter course is taken, however, there is a danger that owing to the collecting of the varnish in depressions, white spots will appear on the specimen.

There are two obvious disadvantages to this method. The decomposed areas discolour and the coating gives an unnatural appearance to some specimens. Unfortunately nothing can be done about it. The choice is that you either lose the specimen now or treat it in this way when it may survive for many years.

The Treatment of Fossils too large to be placed in a Gas Chamber

Many fossils both vertebrate and invertebrate and plant remains are collected on large slabs of rock and in most museums many of these are on permanent exhibition as wall mounts. The removal of such specimens from walls is a complicated process involving the use of several men and there is always the possibility that the specimens may be damaged during the operation. It has therefore been necessary to produce techniques for dealing with this type of fossil. Besides these there are the very large skulls, limbs and girdle bones of giant marine reptiles which cannot be treated in a gas chamber of a practical size.

In the past, solutions of ammonia both in water and in alcohol have been used to treat such objects, but while in many cases they were effective they were extremely unpleasant to use. The operator had to be protected by a respirator and goggles and when the fossil was part of a public display the work had to be carried out during closing hours.

The present-day method employs a 5 per cent (v/v) solution of morpholine in industrial methylated spirit. Morpholine is slightly hygroscopic so the treatment should not be carried out at times when the relative humidity in the region of the specimen is high, for the solution may then take a long time to dry.

When dealing with specimens attached to a wall, the first thing to do is to examine the whole area for loose pieces. Those that can be lifted from the surface without damage should be removed and all decomposition products cleaned off either by brush or needles. The free piece and the region from which it comes are treated with several applications of the morpholine solution, each being allowed to dry before the next is applied. Drying may be assisted with hot-air blowers, provided that there is no build up of alcohol vapour which might be ignited by a spark. Where it is thought unwise to use hot air, drying can be speeded up by blowing normal air over the surfaces concerned with any form of bellows; for small areas, rubber blow-balls, previously mentioned in Chapter 4, may be used.

The loose pieces, when dry, are replaced and stuck with any fast-drying adhesive which is not readily soluble in alcohol. A solution of polymethylmethacrylate in ethyl acetate is one which is effective. Water-based adhesives should be avoided in these cases.

The next step is to examine all cracks both in the fossil and in the matrix. If decomposition products are present in these, they should be cleaned out using dental probes and fine needles. A small pencil-like electric torch and a dental mirror are useful at this stage for it is important that as much of the decomposed material as possible should be removed, and these help the operator to see further into the cavities.

Any area which is considered to be too weak to stand having a solution brushed over it should be strengthened by injecting adhesive into the cracks by means of a polythene pipette or a hypodermic syringe. When all is secure, any surface deposits should be removed from the specimen and the next stage in the treatment can begin.

When there is doubt about the relative humidity being too high a small part containing an area of decomposition should be treated first and observed to see that the solution dries in about an hour. Provided that this first test is satisfactory, the 5 per cent morpholine solution is brushed all over the fossil and the matrix. The whole is left to dry for several hours, when the treatment is repeated. It is important to check after each application that the solution is drying out; as a rule three applications are enough.

The specimen is now left for several days to dry; when all alcohol vapour has dispersed a hot-air blower may with advantage be played over the surface. The whole of the specimen is now coated with a polybutylmethacrylate solution whenever this is available. Most commercial solutions need to be diluted to about one part to three parts of toluene by volume. Where polybutylmethacrylate is not available a solution of 10 per cent (w/v) of polyvinyl acetate in toluene may be used. At least two coats of either is required and care should be taken to see that it enters all cracks and that there are no voids on the surface. Each coat should be dry before the next coat is applied, and the final one must be dry before the next stage of treatment.

Any cracks in the specimen may now be filled with a jute floc-based gap-filling adhesive, or epoxy plastic wood or putty; if nothing else is available plaster of Paris may be used. The filled areas are coloured to match their surroundings, using dry colours and the solution used for coating.

At the end of the treatment there will be red patches on the fossil where the morpholine has reacted with the iron compounds and the coating will have given it a gloss. If it is though desirable, the red patches can be painted out, but their presence indicates that the specimen has suffered decay and therefore that it might happen again and so should be watched.

Large fossils which can be taken to the laboratory for treatment on a bench are easier to deal with. They are much less difficult to handle if they are placed on a paper-covered board rather than directly onto the bench top, for the board can be moved about to facilitate the work without the specimen itself being touched while it is in a fragile state.

The general procedure is the same but in this instance both sides of the specimen can be treated. In extreme cases where decomposition has rendered parts of the fossil untouchable it may be necessary to consolidate it before any neutralising treatment is carried out. A solution of cellulose acetate in acetone may be applied by spray, but the blast must not be directly onto the weak areas. The spray must be arranged so that the jet is either above or to the side of the specimen so that the atomised solution falls as a fine rain upon it. When this has dried and the specimen is more robust, the processes of the removal of deposits, neutralisation with morpholine, repair of loose pieces and cracks, coating and restoration are carried out.

It is once more emphasised that there is no certainty that the morpholine treatment is a permanent cure, but it has prolonged the existence of many fossils and is therefore considered worth doing.

Treatment with Cetrimide and 'Savlon'

These two substances have been tried as treatments for pyrites breakdown. The theory behind the use of both of them is that the trouble is bacterial in origin. 'Savlon' is used as a 10 per cent (v/v) dilution in industrial methylated spirit and cetrimide as 0.1 per cent (w/v) solution in the same solvent.

Experimental use of these is worthwhile, but it seems reasonable to suppose that if they are effective the treatment will have to be repeated at regular intervals. The advantage is that coating with a plastic is not necessary. It is difficult to give an assessment of the worth of these two substances as trial over a very long period is the only way in which this can be proven.

Preservation of Small Fossils by Immersion in Inert Liquids

The idea of keeping small fossils, especially fossil seeds in liquids is not a new one; long ago they were stored under mineral oils or liquid paraffin. Because these were difficult to remove, the practice of using glycerol as a storage medium grew up. The obvious advantage that this could easily be washed off the fossil was also one of its disadvantages for glycerol is highly hygroscopic. This can cause full containers of it to overflow if the stoppers are not airtight. After a period of time the glycerol turns a very dark brown owing to the solution of iron salts from the fossil.

The modern tendency is to keep small fossils under silicone fluid and this is very effective provided that the specimen is dry before it is put into the liquid. If it is not, the water will fall to the bottom of the tube and the fluid will float on it, this results in the fossil being surrounded by water and is very likely to decay. The main disadvantage of silicone fluid is that it is very expensive, a lesser one is that it cannot be removed from the fossil except by the use of organic solvents. That its removal is difficult seems of little importance because the fluid has such excellent optical properties that the specimen may be examined very well while immersed in it.

The transfer of fossils stored in glycerol to silicone fluid is no simple matter, for while from external appearances it seems easy to wash out the glycerol, this is in fact not true. After prolonged washing in a Soxhlet extractor, some small fossil beetles which had been stored in glycerol were dried and appeared to be clear. They were transferred to silicone fluid and two years after they were found to be exuding globules of glycerol. After rewashing and a further period in silicone fluid, some were still not completely free of the glycerol.

Owing to the expense of silicone fluid more work on other possible liquids is needed. It is tempting to reconsider medicinal paraffin and to investigate its effectiveness and the advantages and disadvantages of its use.

Plant Remains from Sea Shores

Fossil cones and seeds collected from the London Clay on seashores are particularly subject to attack by pyrites decay. Fine specimens can be completely destroyed in a very short time after collection and for some reason treatment by ammonia or morpholine and an impervious coating is less successful on these specimens than it is on animal remains. It is possible that collectors do not attempt to wash out the salt or do not do it thoroughly.

Storage in silicone fluid is the most successful means of preservation, but some cones are too large for this to be practical. There has not yet been enough time to tell whether storage in 4-chloro-*m*-cresol vapour is the answer. If it is, palaeobotanists will be well pleased for this method does not involve the coating of the fossil with plastic and as a rule, the application of any form of varnish to fossil plant remains is not regarded with favour. It has, for example, been suggested that fossil cones should be embedded in blocks of transparent plastic and there is little doubt that, if this were done with properly washed and dried specimens, the chances of success would be high, but most scientists working on these fossils do not approve of the idea.

Specimens in Fissile Coals

Fossils from the Kilkenny Coals are good examples of this type. The actual specimens are located on the surface of blocks of coal which may be 15 cm or more in thickness. When the pyrites between the laminations of the coal decomposes, the whole mass is weakened and the surface starts to break up. The cracking of the surface, and therefore the fossil, is often the only outward sign that breakdown is taking place.

The bulk of the coal below the specimen is no longer a source of strength since its weight is a danger to the fossil and is a focus of decay; it is therefore better that it should be removed and replaced by a stable support. The following method has been developed for dealing with this problem.

If any decomposition products can be seen on the surface or in the cracks, they are removed if the condition of the specimen will allow this to be done. Specimens which are cracked but still firm can be cast at this stage if silicone rubber is used, provided that the cracks are filled first. This is most easily done with the water soluble putty based on polyethylene glycol 4000 which has been mentioned before, or some of the molten wax may be used. After the mould has been made, the filling is removed from the cracks and the specimen, according to its size, is treated by exposure to ammonia gas or by application of the 5 per cent (v/v) solution of morpholine in industrial methylated spirit. In this method, the specimen is not oven dried but when morpholine solution is used should be left to dry out for at least twenty four hours.

The next step is to consolidate the surface with an adhesive which remains slightly flexible. A dilution of polyvinvyl acetate emulsion is applied all over the fossil bearing face; this will enter any small cracks and when it has dried the larger cracks are treated with undiluted emulsion. This is best done by brushing it in and any excess which gets onto the areas around the cracks is wiped off with a wet rag before it dries. It should be noted that the emulsion is only applied to the surface bearing the fossil. The emulsion is dried in air for 12 hours and then the specimen is placed in an oven set at 50°C for 2–3 hours. It is removed and allowed to cool. The fossil-bearing face should now be strong enough to be handled.

The removal of the excess coal from the back can now begin. The best tools for this job are thin bladed spatulas of the type used in kitchens or a similar sort which are made for mixing paint. Whenever possible the removal should be done with the fossil uppermost so that it can be constantly observed for signs of movement. A convenient crack in the lamination of the coal is selected and a spatula carefully thrust into it. If the tool is moved from side to side it is possible to feel and to see the line along which the coal will split and the tool can be moved along this line; sometimes the use of two spatulas at the same time is advisable. By these means layers of the unwanted coal can be eased off; at no time should any great force be used. In suitable cases it is possible to reduce the thickness of the coal from several centimetres to about 1.5 cm or less.

There is of course, a limit to the size of specimen which may be treated in this way. Much depends on the condition of the important surface, but in general it is not wise to attempt this process on a block whose greatest length is more than about 25 cm unless a temporary supporting layer is applied to the specimen-bearing surface before the removal of waste material is commenced.

The reduced block must now have a rigid support constructed on its back. The fossil-bearing surface is first screened by cutting a piece of tissue paper to shape and sticking this by its edges with a water-soluble adhesive to the specimen which is now turned fossil side down onto a board padded with plastic foam covered with a sheet of cellophane. A piece of glass fibre chopped strand mat is cut so that it is 2.5 cm wider than the block of coal. Some polyester resin is mixed with the correct amounts of accelerator and catalyst and some of it is painted onto the back of the block. The glass mat is placed on this so that a margin of 2.5 cm stands out all round the block and polyester resin is stippled into the mat, fixing it down to the block. The resin need not be applied, at this stage, to the margins of mat. As soon as the resin has set hard the specimen is turned over onto another board which has been covered with cellophane. The tissue paper is removed from the specimen surface and if any resin has leaked through onto it, it is removed. At this stage it should be possible to cut or scrape it off with a scalpel. Resin may now be applied to the margins of the mat, the resin allowed to harden, the block turned over once more and another layer of glass mat and resin applied on top of the first; this time the glass mat extending over the margin is also treated with resin. When the resin has cured we have the specimen firmly attached to a rigid backing with a stiff margin of the material projecting by 2.5 cm on all sides.

At this stage a decision has to be made as to whether or not the specimen should be washed. It is difficult to give advice on this point as every specimen presents a different problem. The emulsion and the resin should hold most specimens while in water and the author has never lost a specimen as a result of washing.

However there are possible dangers; one is that although soluble iron salts will wash out through the emulsion coat, this may become stained red in places and much time may have to be spent in cleaning it off. The introduction of a small amount of citric acid into the water in the first stages of washing decrease this tendency. If washing is carried out it should be done in repeated changes of de-ionised water until a sample of this gives no white precipitate when tested with a solution of barium chloride. The water in the last stages must not contain citric acid, nor must this be used at any stage if the fossils are calcareous. The best policy to adopt in the decision to wash or not is, 'When in doubt, don't'.

When a specimen has been washed it should be dried slowly by wrapping it in a 6 mm thick layer of surgical cellulose soaked in de-ionised water and letting it stand at room temperature until dry. The wet pack should be in contact with every part of the surface of the specimen so that as the water evaporates from the surgical cellulose, it will draw any soluble salts out of the specimen and deposit them on the outside of the pack. On removal of the cellulose the specimen should be oven dried at 50°C for some hours and when it is cool the edges are built up on the margins of glass fibre to the level of the specimen with a mixture of torn up surgical cellulose, or some form of floc and polyester resin having a consistency which makes it spreadable with a spatula. It may be built up in several layers, and when it has been left to harden for four days it may be trimmed to shape and smoothed. An easy way of doing this is to use 'Abracaps'* or some other form of small grinding tool which fits into a heavy duty handpiece made for dental mechanics. This work should be done under a dust extraction hood and a mask should be worn.

The smoothed edging may now be painted to match the coal. The next stage is to coat the fossil and the matrix with either a solution of polyvinyl acetate in toluene or with 'Bedacryl 122X'.

The finished specimen is enclosed in a watertight and airtight box and the exposed surface coated with an impervious varnish. In this condition its chances of survival are good since it has also lost some excess weight, its edges are protected and it is easier to store.

From the description of this technique, it is obvious that it is one which has to be carried out with the greatest care. It is advisable, before attempting it on a specimen, to carry out experiments on some spare pieces of the matrix to find out just how easily the coal will split and what happens to it if it is immersed in water for at least a week. The polyester resin used should be one having a high resistance to water and should be made fire resistant by the inclusion of a fire retardant recommended by the makers. 'Trylon W.R. 180' with 2 per cent accelerator and 2 per cent catalyst, with 15 per cent of 'Prefil F' will meet these requirements.

Possible Future Approaches to the Pyrites Problem

It has been noticed that fossils containing iron pyrites are most likely to decay if they are kept in an environment in which the temperature and relative humidity of the air are subject to great

---

* The trade name for hollow cylinders or cones of stiffened and very durable paper or cloth charged with an abrasive. The base is open and fits over a solid rubber cylinder or cone. The rubber carrier is attached to a spindle which fits into a rotary handpiece at the end of a flexible arm and is driven by a motor. The cylinders, cones and rubber supports are available in a wide range of sizes.

variations. This would seem to suggest that there are optimum conditions for these two factors at which the specimen should be stored. Long-term experiments, using cabinets in which the temperature and humidity can be varied at will, might help to establish whether or not this is the case.

There is room for a great deal more research to be done into the problem in general. Unfortunately a very long time has to elapse after a treatment has been carried out before it can be said to have been successful. As a result of research done by archaeological workers on building stone, the use of either barium hydroxide or barium chloride has been considered and a few specimens have been treated with these substances, but not long enough ago to make it possible to discern whether this is an effective treatment. Anyone trying this technique should remember that soluble salts of barium are poisonous and great care will have to be taken to see that none remains on or in the specimen. The theory is that the sulphate formed by decomposition would be converted to barium sulphate which is stable. If barium chloride was used, any chlorides resulting from the reaction would have to be washed out as well as the excess of the barium salt itself.

References

Bannister, F. A. (1933) 'The preservation of pyrites and marcasite', Mus. Journ., 33, 72–5.

Booth, G. H. and Sefton, G. V. (1970) 'Vapour phase inhibition of thiobacilli and ferrobacilli. A potential preservative for pyritic museum specimens', Nature, Lond., 226, 185–6.

Radley, E. G. (1929) 'The preservation of pyritised and other fossils', The Naturalist, April 143–5; May 167–73; June 196–202.

8
Mounting Fossils for Exhibition

Specimens on Horizontal Surfaces

Fossils which are to be displayed normally require to be provided with some sort of support to hold them in position. In the simplest case, that of moderately heavy flat-bottomed specimens which are to be shown in a case with a horizontal floor, all that is required is some means of preventing them from moving as a result of floor vibration. Modern exhibition techniques very often employ coarse woven cloth as a background material, and this attached to the floor of the case is usually enough to hold this type of exhibit in position; should it not be, then a few discreetly placed headless pins will cure the trouble.

Small specimens whose shape makes them inclined to roll, such as brachiopods and molluscs, need to be anchored in some way even when exhibited in horizontally-floored cases and there are numerous widely-used ways of doing this. Normally the fossil is not placed directly onto the bottom of the case but is first mounted on a rectangular support; wooden tablets about 6 mm thick and covered on the upper side with thick paper, or pieces of coloured cardboard are to be seen in most museums fulfilling this role.

Methods of holding the specimen to its tablet vary from placing headless pins round the fossil to sticking it down using a plug of cotton wool (cotton) soaked in a glue which is easily soluble in water. When pins are used they should be longer than required and having been driven in, are then cut off. This method is not to be recommended, if it can be avoided, as the pins have to be placed so that they hold the specimen tightly and there is a risk of damaging it. The placing and removal of a specimen which is stuck to the tablet by cotton wool and adhesive is made easier if a hole is bored through the support and the adhesive plug projects slightly into it. The unfilled part of the hole gives access to the back of the plug and the specimen may be removed by placing a wad of cotton wool (cotton) soaked in warm water into the hole.

Specimens may be held steady in horizontal position if they are placed in a cavity which conforms roughly to the contours of the under side. This has been done by casting blocks of plaster and, when it has partially set, pressing the fossil, which has been previously lubricated with olive oil, into it. The fossil is then removed from the fully set block and the surface of this rubbed down with fine glass paper, after which it may be painted. At first sight this seems to be a simple method, but in practice it is not, for to arrive at a result which is acceptable requires a great deal of time and skill. It is almost as quick to cut a cavity into a block of soft wood with wood carver's gouges, or to cut a hole with a fretsaw and then shape the edges with small files.

A dress-making material called 'Velcro' may be used to hold some fossils in position. The 'Velcro' consists of two strips of fabric, on one there are a large number of very small plastic hooks and on the other a large number of small plastic loops; so that when the two sections are pressed together the hooks and loops engage and hold fast. If one piece is stuck to the mounting board and the other to the back of a fossil, the specimen will remain where it is put until deliberately pulled off the support. The removal and replacement of objects held in this way is extremely easy and the link between the two parts of the 'Velcro' may be made and broken hundreds of times before it shows any signs of weakening.

Expanded polystyrene tiles, of the type much used in internal decoration, are a good backing material for exhibits. They are cut easily with sharp tools, or with a special battery-operated hot wire (see Appendix I). Specimens may be stuck to them or recessed into them. Details for working the material are discussed in Appendix I.

Specimens on Vertical or Sloped Surfaces

The majority of exhibits of fossils are assembled on vertical or steeply sloped surfaces. Sizer (1960) published a method for mounting specimens in these conditions which is so simple and so elegant that it should be considered before any others. He attached bolts or wires to the back of the specimen with the jute floc compound 'Fibrenyl'. These were then passed through holes drilled in the back board, and in the case of bolts, were secured by nuts and where wires were employed these were bent over. The wire method is particularly good if peg board is used. When the exhibit is dismantled the bolts and wires can be removed by softening the 'Fibrenyl' with industrial methylated spirit, acetone or ethyl acetate. This method as well as being very simple and effective has the added attraction that no part of the supporting bolts and wires are visible to a person looking at the exhibit.

There are many fossils to which, owing to their shape and weight, Sizer's method cannot be applied and for these, mounts have to be made from plastics or metals.

'Perspex' mounts may be seen in many places, and more often than not the plastic is used as though it were a substitute for wood and is shaped by cutting and joined with screws.

Claringbull (1948) described several ways of making mounts for minerals in which the fact that 'Perspex' becomes malleable at about 80°C is used and the methods described are readily applicable to fossils. The most useful type of mount is one which has four arms which are shaped to the specimen and grip at four points on its edge. Almost any shaped object may be held in this way. Time and materials are saved if the mount is first fashioned in thin card, and this used as a pattern to cut the required shape from plastic; this is normally a bar with four arms projecting from it. The plastic will soften and become easy to bend if it is passed rapidly back and forth through a small flame. If the barrel is unscrewed from a bunsen burner a nipple with a small hole in it will be found through which the gas passes and this provides a suitable flame. Care should be taken not to overheat the plastic or it will burn. The arms may be bent in stages to fit the specimen or they may be shaped by pressing them, while still hot, onto the fossil. In either case the formed plastic must be held in position until it has cooled enough to become rigid again; otherwise as soon as pressure is taken off, it will revert to its original shape. In the malleable state 'Perspex' is too hot to be worked with the bare hands and if one wishes to form it with the fingers, they must be protected by a glove. No great force is required to bend this material and light tools like forceps may be used; the type of wooden clothes' peg which is closed by a spring may be converted to a useful tool for this purpose by grinding down the sides of the gripping end so that it takes on the shape of the jaws of a pair of pliers (Fig. 11). Mistakes in bending may be corrected by reheating the plastic at the point at which the error has occurred.

Fig. 11. Clothes peg adapted for bending plastic sheet.

When the mount fits the fossil, the four ends of the arms of plastic which appear on the displayed side as small lugs, may be rounded off and shaped with fine glass paper and re-polished with a soft cloth and metal polish or by buffing on a fine cloth wheel. The plastic is flexible and at least two of the arms of the mount should be capable of being pulled slightly apart to allow the fossil to be placed on and taken off the mount while this is in position in the case. The mount is fixed to the case with a countersunk screw through the main bar of plastic from which the four arms originate.

'Perspex' will weld to itself if the two surfaces to be joined are grease free. Junction is achieved by clamping the two pieces together using clothes' pegs, rubber bands or weights, and running chloroform from a brush round the edges of the top piece where it contacts the bottom one. The liquid will be drawn into the joint and within a half an hour or less a strong weld will be formed. This makes it possible to build up quite complicated mounts from strips of the plastic.

Cellulose acetate sheet may be used in place of 'Perspex' over which it has some advantages; it is cheaper and more flexible, which makes it easier to cut. This can be done by scoring it with a sharp point and snapping the pieces apart. The same method may be used with 'Perspex' but it requires more care. Cellulose acetate is not water white as is 'Perspex', but has a slightly blue tinge. It may be welded in the same way as 'Perspex', but acetone or methyl ethyl ketone is used in place of chloroform. The addition of 5 per cent v/v diacetone alcohol to either solvent slows its evaporation; this is an advantage when making large welds. There is a limit to the thickness of either plastic which is used for mounts; for if it is above 3 mm the small lugs which appear on the surface of the specimen reflect so much light that attention is drawn to them rather than to the specimen displayed.

Owing to the weight of a specimen or because a plastic mount is too obvious, the traditional method of using metals often has to be resorted to, the most widely-used materials being iron, tin plate and brass. All of these will support a specimen weighing several ounces if they are about 1.5 mm thick. Iron and brass may be bought in long strips approximately 1.5 cm wide, and from these pieces may be cut to make supports for a wide range of fossils. The length of metal can be estimated by placing pieces of tape or string on the fossil along the lines on which the supporting arms are to run; it helps if these are first drawn on the specimen in pencil. Slightly more than the estimated length should be cut from the main strip and at least 2.5 cm will be required to take one or more screws for fixing to the display board.

When iron is used the arms of the mount are formed from the main strip by making the necessary cuts into its length to provide the number required; a mount with four arms requires three equally spaced cuts. The cutting is done with a hacksaw fitted with a fine-toothed blade.

The section in which there are no slits is held in a pair of pliers and the ends of the separated pieces adjacent to it are made red hot and bent away from each other into the relative positions they will occupy in the finished article. At this stage the cut edges should be smoothed and slightly rounded with a fine cut file so that there is no danger of their abrading the specimen, and then the arms are bent with pliers to conform as closely as possible to the contours of the specimen. It is best to do this to each arm separately, fitting one before the next is started. The ends of the metal which show on the front of the fossil may be thinned both to

158 *Mounting Fossils for Exhibition*

make them as inconspicuous as possible and so that the sharp bends needed at these parts are easier to make. Apart from the initial separation of the arms mentioned above it is not, as a rule, necessary to heat the iron in order to bend it; using two pairs of pliers, metal of this thickness can be worked cold (Fig. 12).

Brass is worked in the same way, but after the necessary cutting has been done, the metal needs to be softened by making it red hot and plunging it into boiling water. When it has cooled it can be worked cold. This metal also has the attraction that pieces can be soldered together; this means that the arms of the mount can be made of separate pieces soldered to the main support, and they do not all have to be of the same thickness. This applies equally to tin plate. Resin-cored solder is better than solder and an acid flux. All that is necessary to make a good joint is to clean both surfaces to be joined to brightness with fine glass paper and then to apply a thin layer of solder to each with a well tinned soldering iron. If the two solder-covered faces are placed together and held in position with a piece of wood, and the iron is applied to the top piece the solder will flow and, provided that pressure is maintained until the

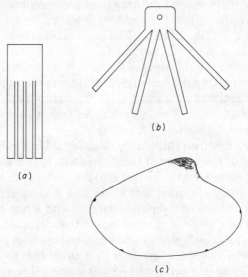

Fig. 12. Three stages in making a claw mount.

Specimens on vertical or sloped surfaces 159

metal cools, a strong joint will result. As the iron is removed pressure is kept up by the piece of wood used to steady the joint. When joints are small enough, they may be held together with a wooden clothes peg before the iron is applied and the peg left in position until the solder cools.

Metal mounts, whenever possible, should be designed so that the piece that is screwed to the backboard of the display is hidden by the specimen when this is in place. It follows that the supporting arms must be positioned in such a way that the fossil may be placed on the mount after this has been screwed into position.

The type of 'claw' mount described above cannot be used, with ease, on specimens which are too large to fit into 5 cm square or project more than about 4 cm from the mounting board. Larger specimens are best supported by a series of separate clips placed along their length and breadth; these are much simpler to make than the 'claw' mounts. The one essential thing to remember, when mounting a specimen in this way, is that each clip must be returned to exactly the same place every time it is removed. To ensure that this happens the making of the clips should be carried out with the specimen placed on a convenient sized board to which a piece of stiff card has been pinned by its corners. The horizontal and vertical axes are marked on the card in pencil and the specimen then placed so that its axes correspond with these; then a pencil is run round the edge of the specimen, thus drawing its outline on the card. If the underside of the fossil is not flat and it rocks on the board it can be steadied by placing wedges and blocks of wood or polystyrene foam below it. These may be stuck to the back of the specimen if it is in a block of matrix with nothing of interest on the underside, otherwise their exact positions must be marked on the card by drawing round them with a pencil. They can be conveniently held in place with polyethylene glycol 4000 applied hot.

With the fossil correctly positioned, a clip is formed to fit the lower edge by bending the metal into a right angle at a point which leaves a tang of sufficient length to accommodate two holes for wood screws. The holes are drilled and the long arm placed so that it touches the side of the specimen and the angle of the metal is as close as possible to its base. Pencil marks are made on the card through the drilled holes, to ensure accuracy in inserting the screws. Another pencil mark is made on the metal at the point level with the top face of the specimen; the metal is removed from the board and bent over at this mark to form a small lug which will hook over the edge of the fossil. This lug is shaped, filed and finished off neatly to render it unobtrusive and the clip screwed back into position.

The next clip on this edge is made and fastened in position. A light pencil line drawn round the parts that overlap the surface of the specimen records the point at which they are placed. Two clips on the lower edge are usually enough, but others may be needed if the specimen is long or if cracks are present. Normally a fossil will be adequately supported if two further clips are made to fit it one at each side. The transfer of the mount to the case is made by using the card, on which the outlines of the specimen and the parts of the clips through which the screws pass have been marked, as a template. If the blocks have been stuck to this, they are removed and their marked outlines cut out. The card is placed on the backboard of the exhibit and its horizontal and vertical axes lined up with those of the case. The positions of the screws for the clips and of any supporting blocks are marked by drawing round the holes in the card. This method can be used to mount very large and heavy specimens, but this involves working with mild steel of 6 mm or more in thickness. Steel of this size obviously cannot be worked cold but has to be constantly heated and bent on anvils or in a vice, so unless the preparator has had some training in metalwork of this type, it is better to employ a blacksmith.

Vertebrate Skulls

Skulls of reptiles and mammals vary so greatly in size, weight, depth of palate and types of tooth row that it is only possible to offer some general hints on making mounts for these. In almost every case it is desirable to have a central bar running from the front of the palate to the occipital region, to this are attached, more or less at right angles, pieces which stop the skull rocking from side to side and hold it in position on the mount. The exact placing of these is determined by the type of skull but they should give support as near to the occiput and centre of the palate as possible. To prevent the skull slipping off the mount they need,

Vertebrate skulls 161

where possible, to be shaped so that they overlap and sometimes grip the sides. The rear one can usually be made to do this either just behind or just in front of the root of the zygomatic arch but care must be taken not to foul the area of the joint between the upper and lower jaws. The forward support is more difficult. If there is a long diastema in the tooth row, advantage may be taken of this, but in its absence it is necessary to place the holding clips behind the tooth row and to support the palate at its centre by a crossbar which stops short of the sockets of the teeth (Fig. 13).

Some skulls have deep narrow palates and it is not safe to run

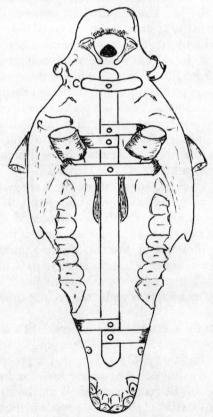

Fig. 13. Palatal view of an equine skull showing possible positions for supporting clips.

an unpadded metal bar into these as the edges will damage the bone. For large heavy skulls of this type, it used to be the practice to carve wooden blocks so that they exactly fitted the bone, and these were screwed to the central bar. It may be still necessary to do this sometimes, but when weight will allow it, the possibility of making a mould of the area in polyurethane foam or some other material should be considered, as this will save time. The main consideration is that the weight of the skull must not crush the material and cause it to expand sideways as this would, in some cases, break the bones of the palate. There is no such problem in dealing with light skulls which have deep palates. Provided that they will stand the pressure required the process is simple. A mould may be made of them in a soft polyvinyl chloride impression compound, like Vinagel 116, which can be heat cured. The palate is dusted with talc and a piece of the compound is gently pressed into it. A length of mild steel bar of the size selected for the support may be pushed into the plastic and the edges of the latter tidied up. The metal, with the plastic mould on its end, is carefully removed and placed in an oven at 150°C until the plastic is fully set. This usually does away with any need for clips or any other support at the front of the skull. Several human and ape skulls have been mounted using this technique.

In the human skull and some others, the foramen magnum is underneath and advantage can be taken of this to reduce the amount of metal which is visible on a mounted specimen. A suitably-sized cork can be fixed to the central bar of the mount and pushed into the foramen, or alternatively two prongs of metal covered with plastic tube may be positioned to pass into it. Either of these means will stop sideways movement and it is then only necessary to provide a support under the occiput to stop the skull rocking.

The shaping of the metal to fit the underside of a skull is much easier if the skull is supported, palate side up, in a box of sand or in some other convenient way. The old method of propping it up in the position it is to assume on the finished mount and working underneath it is not to be recommended. With the palate uppermost, the constant testing of the fit of the metal to the bone is simple. The central bar is shaped first followed by the side clips and cross supports. Depending on whether these are fixtures into which the skull may be rested or whether they are clips to hold it in position which must be removable, they are fixed to the centre bar either by riveting or by screws which pass through the clips and drive into tapped holes in the main member.

The finished mount will be either attached to a vertical or sloped surface or it will stand on the floor of a case or on a shelf. The support will be either a bracket passing from the central bar of the mount, under the skull and onto a surface behind it, or an upright rod. In the case of the bracket, the lower jaw should be placed in position and the bracket shaped so that it does not interfere with the articulation. When lower jaws are placed on top of upper jaws, it is of vital importance to put some soft material between the tooth rows so that one may not damage the other. The jaw does not need to stay in position while the bracket is fitted, in fact it is far safer if it is removed as soon as the correct positioning of the bracket has been determined. This having been done, it can be replaced and a fitting made to suspend it from the central bar. The form of suspension depends on the length of the jaw. Short jaws can usually be held by three hooks, one being the end of a bar brought down from the central support which fits under the chin and the other two being attached to it and fitting just in front of the ascending rami. Long jaws require more complicated supports which may have to include two hooks and two locking clips for each ramus.

Skulls standing on floors or shelves are supported by one or more upright rods. These must be at right angles to the floor but are often not at right angles to the palate of the specimen. When a single support is used, this is best placed at the point of balance of the specimen so that the weight is distributed evenly on all sides and the rod does not tend to bend, and as little strain as possible is put on the bolts which attach it to the rest of the mount. The angle at which the bar running along the base of the skull has to meet the upright rod can be best arrived at by making a sketch of the skull in the required position to the horizontal and then drawing a line representing the supporting rod which is at right angles to the horizontal and meets the skull at the desired place. The angle can then be measured. Another way is to place the skull at the required angle, palate side up in a sand tray, with the parts of the mount already made in position. If a long spirit level is now placed with one end on the highest point of the front of skull and adjusted until the bubble is central, the box of the instrument will indicate the position of the floor of the case relative to the angle of the skull. If no assistant is available, the level should be clamped in position with retort stands. The supporting rod can now be made using this line to check that it is upright (Fig. 14).

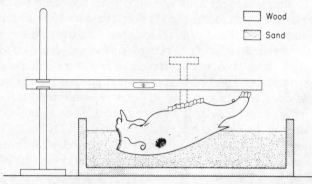

Fig. 14. Use of spirit level to ascertain the angular relationship between the upright and palatal supports in a skull mount.

The support may be either a solid rod or a tube. For heavy specimens, tubes are better since diameter for diameter they are more rigid than rods. When rod is used, it may be attached to the main mount by making a cut through the diameter of sufficient length that when opened, two arms of the required size are formed. The iron needs to be red hot when the cut is opened. The two arms so formed are bent until they fit the bar of metal running along the bottom of the skull and the main shaft of the rod is at right angles to the horizontal. The rod may be fixed to the mount by screws passing through the arms and tapped into the central bar or in some cases by rivets. If a tube is used, the end which touches the main mount is filed so that it sits as flat as possible upon it, and is at right angles to the horizontal. It may now be clamped in this position and two brackets made so that one leg of each fits the central bar and the other the side of the tube; these are fixed to the main mount and to the tube by screws in the usual way.

When fitting one piece of metal to another, it is necessary, at an early stage, to mark the position of the one being bent, upon the one which it is to fit by drawing in pencil the outline of the end and sides of the former on the latter, so ensuring that after every adjustment the arm goes back into the right place. After a satisfactory fit has been made holes are drilled in the piece that was being manipulated, which must be large enough to give clearance to a suitable size metal screw. This piece of metal is now placed on the other so that it fits its own outline which has been previously drawn, and the point of a sharp pencil is placed through the screw holes and their positions marked on the other piece of metal. These pencil marks are used to indicate where the metal should be centre-punched and then drilled with holes which are the tapping size for the selected screw. These holes are then tapped to form the thread into which the screw is driven.

The length of the upright in this type of mount may depend on the design of the exhibit, but in general it should be as short as is consistent with the easy assembly of the mount when it is in its final position. Long supporting rods, especially in the case of human or hominid skulls, can convey an unfortunate impression of a cocoanut shy.

The fixing of the rods to the bottom of the case, to a shelf, or a wooden block is most easily done by attaching a metal flange to the lower end and drilling this to allow wood screws to pass through it into the base. The type of flange used in gas fittings is excellent for this purpose, but a set of gas fitter's dies are required to make the threads on the end of the rods or tubes. Small skulls may be fixed by threading the end of the rod to take a hexagonal nut, for a length which will allow one nut to be screwed up to the top of the thread and leave enough to pass through a hole in the wooden base and for a second nut to be screwed on below. In the case of very light specimens a threaded rod may be screwed straight into the wood, or without threading, it may be pushed into a hole and held in place with a little adhesive.

Isolated Limbs

The method for mounting vertebrate limbs naturally depends on their size, but in every case the starting point is the manus or pes.

166 Mounting Fossils for Exhibition

Fossil bone should not be defaced by drilling holes in it so the methods of the osteologist cannot be used. As a rule there is no objection to sticking the bones of the tarsus or carpus together. Polyvinyl acetate emulsion has advantages, when the bones are light; it is thick enough to fill small gaps and when it is dry looks very much like cartilage; a jute floc gap-filling adhesive will hold all such bones but the heaviest, and the assembled carpus or tarsus may now be treated as one element. It is now necessary to position the metapodials and phalanges. The articulation of the proximal ends of the first with the tarsus or carpus is obvious and they may be held in position in several ways. Once more, if the bones are light they may, with advantage, be stuck in place.

In cases where this is neither possible nor desirable, the podials, metapodials and phalanges may be supported at the back by setting them into a block of plasticine or modelling clay (Fig. 15a). The former is easier to use as no precautions have to be taken to stop it drying out. The areas of bone which will come into contact with the modelling material should be lubricated with a solution of soft soap, which is allowed to dry before proceeding with the work. If it is intended, eventually, to support the digits on strips of metal, the only thing that matters at this stage is that each bone is correctly articulated and the face of the plasticine or clay need not be neatly finished, but there is a way of mounting the manus or pes in a block of plaster of Paris and in this case the modelling material needs to be smoothed and finished accurately because it is cast to make the final support.

The hand or foot is assembled from the ungual phalanges upwards and each bone is set in the clay so that it is in a depression whose sides will hold it in place but from which it can be lifted (Fig. 15a).

When this has been done, the bones may be removed and the surface of the modelling medium smoothed on the front and sides (Fig. 15b). From this a plaster mould is made (Fig. 15c). Where plasticine is used it is first given a coat of olive oil so that it will separate easily from the set plaster which should be slightly tinted by mixing a little yellow ochre powder into it in the dry state. The plaster is mixed by sprinkling it into a vessel of water, stirring it to a lump-free paste and then allowing it to stand until it has thickened to the point where it may be picked up on a spatula and

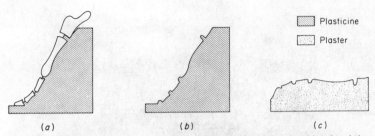

Fig. 15. Stages in mounting a mammalian foot. (a) Mounted in plasticine; (b) Bones removed leaving plasticine mould; (c) Plaster cast from (b) which may be used to make plaster or plastic support for the foot or as a template to make a metal mount. (All bones stylised only one digit shown and the tarsus represented as a single block.)

spread over the front and sides of the clay. The plaster must be pushed into the depressions, but care should be taken not to distort them by allowing the spatula to press upon them. The thickness of plaster spread should not be greater than that which will make the set mass strong enough to be handled without fear of its breaking. It should be left to harden for at least two hours, when it may be removed from the modelling material. The cast so formed is the reverse of the mould and the projections represent the backs of the bones.

If this cast is now taken and given several coats of a solution of soft soap and this is allowed to dry, a plaster reproduction of the original surface of the clay or plasticine may be made from it. This time the plaster used should not be tinted. It may be spread upon the first cast, as described above, or it may be poured on to it in a more liquid state if a retaining wall of plasticine is first built. The thickness of plaster applied should be just a little more than is required for the finished mount.

When this batch of plaster has set and become hard enough to handle, it may be separated from the other; in most cases it will pull off, but if much resistance is felt when this is attempted, the tinted plaster must be cut away from the white.

The mount is now set aside to dry out and may then be smoothed where necessary, if the original component has been neatly finished, very little work should be needed anywhere but on the back.

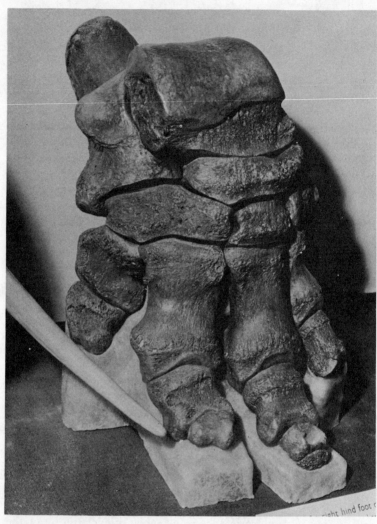

Plate 13. Aveley Elephant (3): Right pes after laboratory treatment, mounted for exhibition on a fibreglass-resin support. (By permission of The Trustees of the British Museum (Natural History).)

Fibreglass may be used in place of the white plaster for the end-product. The yellow plaster impression must be dried and shellaced and given a good coating of wax polish. The first layer of mat used should be a fine surfacing type, the second ordinary chopped strand mat and the back should be finished with surfacing mat.

Successful mounts for elephant feet can be made by using the original clay or plasticine support as a male mould and building up fibreglass on top of it; but since, in this method of application, every layer makes the depressions in the final cast shallower, the number of glass layers should be as few as possible.

The modelling material is removed from the inside of the cast when the resin has set and the bones can be secured in place by fixing bolts or wires to their backs, which finally pass through holes in the fibreglass and are fixed either by nuts or bending.

Plaster or fibreglass mounts for manus and pes are acceptable in many cases, but in others the appearance of the finished mount is improved if the hand and foot bones are supported on metal strips. When making such a mount the plaster impression taken from the plasticine may be used as a pattern and the metal formed to fit the casts of the backs of the bones which are represented. A narrow metal bar is bent to fit each digit and is continued upwards onto the tarsus or carpus; as each bar is finished it may be temporarily cemented to the cast with polyethylene glycol 4000. After a support has been made for every digit two bars are fashioned in the area of the tarsus or carpus which touch a part of each of the supports for the digits. Holes are drilled at these points and their counterparts marked on the digit supports. By drilling holes at these points, which may either be tapped to take small screws or made large enough to take rivets, the complete mount for the manus or pes may be assembled into a rigid entity. The bones making up the hand or foot may be held in position by small clips or by attaching wires or bolts to their backs and fastening these through holes drilled in the metal. The assembly will be finally attached to the main support; since it carries little weight it may be made of very light metal.

To continue with the mounting of a limb; the assembled manus or pes, on its support, is placed on a wooden board and where necessary held in the correct position by pieces of plasticine

170 Mounting Fossils for Exhibition

placed at the sides. The lower limb bones, tibia and fibula or radius and ulna are now placed in correct articulation and in all but very large animals can be held in place by clamping in retort stands or by clamps or adhesive tape to laboratory scaffolding. In many cases the radius may be stuck to the ulna and the fibula to the tibia. The femur or humerus are then placed in position and held in the same way. Their support must be arranged so that the backs of the limbs are accessible and there is room to manipulate the metal which will form the main support (Fig. 16). Some very heavy specimens may have to be held on wooden stocks fashioned to fit the front of the bones which are strapped to them by metal bands. While the wooden structure is being made, the bones may be suspended by ropes from overhead beams.

When all the elements of the limb are in the correct position a chalk line is drawn down the back of them indicating the intended

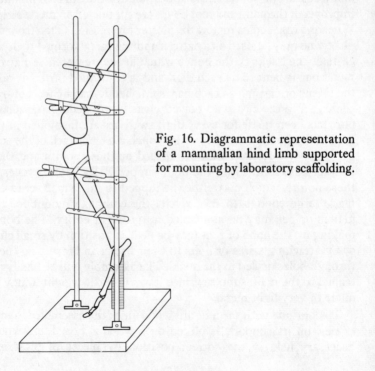

Fig. 16. Diagrammatic representation of a mammalian hind limb supported for mounting by laboratory scaffolding.

run of the main support. The metal is fitted from the ground upwards. The first thing to decide, therefore, is how it is to be attached to the base. Very often a length of the bar bent at an angle at the end and drilled to take wood screws is all that is required; in other cases, a metal plate may have to be welded or brazed onto one end. The first method, apart from being the simpler, has the attraction that the end of the support can be hidden under the animal's hand or foot. Having decided which method is to be used, the mild steel should be bent until the lower part which stands on the base can be placed in the position required. The holes for the screws are drilled in the metal and in the wooden base. The position is marked on the board by drawing round the end and sides in pencil. The hand or foot is temporarily removed to do this. The mild steel is now bent to fit the bones along the chalk line which has been drawn (Fig. 17a–d).

This type of fitting is best done by marking the metal with chalk at the point at which it is to be bent and clamping it at this point in a metal vice. The bending may be done either with adjustable spanners of various sizes or by placing one end of a bar of metal, which is wider than that being bent, just above the chalk mark and hammering the other end. All but the sharpest bend can be made cold. The making of main supports is a tedious business as it means that the metal has to be repeatedly tried against the bones and, at least in the final stages, the screws securing it to the base have to be removed and replaced every time this is done. In some cases it is possible to make a pattern for the main support in the square section aluminium bar sold for making armatures for clay models. This is very easily bent and the mild steel can be fitted to the pattern rather than to the bone.

When the main support fits satisfactorily it is screwed down and such clips as are needed to hold the bones to it are made. These must be fastened either to the back or on the sides of the main support, depending on its thickness (Fig. 18). If it is thick enough to have holes bored into its sides which can be tapped to take the right size screws, the fitting of the clips is slightly easier. Clips should be made of the thinnest possible metal consistent with the safety of the specimen; this makes the work easier and the clips on the finished mount less obvious.

It is, once more, important to mark the position at which the

172 *Mounting Fossils for Exhibition*

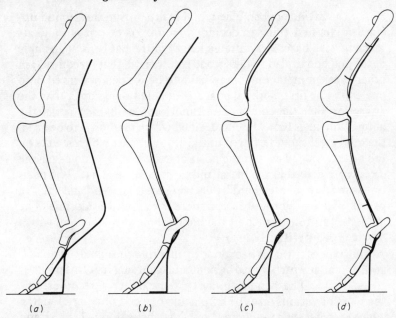

Fig. 17. Four stages in the fitting of the metal support to the limb in Fig. 16. (*a*) Metal bent to fit base board and to meet the foot near the top of the metatarsals; (*b*) Bending continued to fit tarsus and tibia (N.B. metal passes behind the calcaneum); (*c*) Bending continued to fit femur (N.B. metal passes between distal condyles); (*d*) Positions of the supporting clips.

clip is to be fixed to the main support and to make sure that at every test of the fit it is correctly positioned. Four clips are generally needed to hold each long bone, two near the proximal and two near the distal end (Fig. 17*d*). Heavy bones may need a support placed under the distal ends to stop them dragging on the side clips and being damaged. As each bone is fixed to the main support by its clips the temporary support may be removed.

The metal work of the completed mount should be neatly finished by filing the ends of the clips and removing all sharp edges. If necessary, the inside faces of the clips may be lined with soft cloth to stop them cutting into the specimen. The final step is to paint the metal in a colour which makes it as inconspicuous as

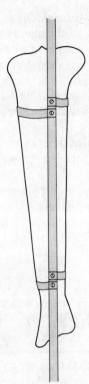

Fig. 18. Supporting clip fastened to the back of a main iron.

possible; the parts which are visible on the bones are made less so if they are painted to blend in with the colour of the specimen. The main object of mounting any specimen is to show it to its best advantage and so the less the metal work shows the better, and this should be borne in mind at all stages of the job.

Mounting Complete Skeletons

The skeletons of fossil animals vary enormously in size and condition and it is therefore only possible here to offer general hints on how to mount a complete skeleton. The first thing to be done, whatever the size, is to decide the pose. There may not be a free choice in this as the possibilities can be limited by factors inherent in the specimen in question; some of the vertebrae may be joined together by matrix and this limits the curvature which can be put in the spinal column and a similar condition of the manus and pes can reduce the number of positions in which these may be placed. Within these unavoidable restrictions, the aim should be to make the posture of the skeleton as interesting as possible. Sometimes it is only possible to mount a quadruped with its four limbs placed so that manus is in line with manus and pes with pes from side to side, and the head looks straight in front of it with its jaws closed. Such a specimen is not inspiring and several of them in one gallery make a very boring exhibition. The appearance is improved if the bones are mounted so that the limbs are in a walking position, and even turning the head very slightly to the side makes a great difference. When there are living animals, which are anatomically comparable with the fossil, photographs of them in motion are invaluable in deciding how the specimen should be posed.

Procedure for Mounting a Quadruped

The pose having been decided, the work may be carried out in the following manner.

Stage 1. The Vertebral Column

The vertebrae are arranged on a board so that they are in correct articulation and follow the curve dictated by the pose. The simplest way to support them is to place them on their sides on blocks of plasticine which are high enough for the tips of the longest transverse processes to be clear of the board; it is important to check that each joint is anatomically possible and that all the neural spines are in line with each other.

If the animal is small or two people are working on the mount, the steel support for the spine may be fitted by bending it to the curve of the underside of the column and testing the fit against the bones, but if the specimen is large or one man is working alone it is better to arrange the vertebrae on their plasticine blocks on a sheet of hardboard and by using a carpenter's square, mark the lowest anterior and posterior points of the centrum of each vertebra on the board. When these points are joined a curve is formed which corresponds to that of the spinal column and if the hardboard is cut along this line; the piece on which the vertebrae were standing may be used as a template and the metal bar fitted to this. The position of the atlas vertebra and the sacrum should be marked on the template.

The column may be supported in one of two ways; a mild steel bar may be bent to fit the curve formed by the undersides of the vertebrae which are then fixed to it by clips, or a bar may be run through the neural arches. The first method is the easier and has the advantage that individual parts of the column can be removed for examination with very little trouble and for this reason will be described in detail.

The linear equivalent of the curve of the column is found by measurement of either the articulated bones or of the template and to this measurement is added the length of the skull. A mild steel bar of suitable width and thickness is cut to this length and a point equal to the length of the skull is marked off from one end. This line indicates the joint between the atlas vertebra and the condyles of the skull.

It is obvious that the width and thickness of the bar depends on the size and weight of the vertebrae. The thickness should be the least that will support the weight of the column without bending. In the finished mount the vertebral support is normally rigidly attached to two upright supports, and this should be taken into account when selecting the material since a bar which is so fixed and has a slight upward curvature will carry more weight without bending than it would if it were straight.

Using the line marking the position of the condyles and the articular facets of the atlas vertebra as a reference point, the bar is now bent to fit the whole length of the column. The bending is done a little at a time and the fit is constantly checked. The gauge of metal required for anything but the largest animal can be bent cold. This can be done by marking the place at which the bend has to be made and then lightly clamping the bar in a heavy vice and pulling it in the right direction with the hands. For sharper bends it should be clamped tighter and long handled adjustable spanners used to obtain the necessary force. Heavier metal has to be made red hot and bent either in a vice or by hammering on an anvil. When a satisfactory fit has been achieved the bent bar is held in its correct position in contact with the vertebrae or template and the position of the back of the sacrum is scribed into the metal, which is then set aside and the vertebrae, when a template is not used, removed from the plasticine supports.

The metal support is now placed with the side which fits the ventral aspect of the vertebrae uppermost and is held in position by retort stands, laboratory scaffolding and clamps or by any other means which will keep it stable. Because the post-zygapophyses of a vertebra fit over the pre-zygapophyses of the vertebra behind it the assembly of these bones on the back support is started at the sacrum and proceeds forward from this point. First the sacrum is placed on the bar so that its posterior end coincides with the mark indicating this point with the mid-line of the body of the bone central to the support and the dorsal spine or ridge upright. Depending on its size the bone may be held in position with pieces of plasticine, rubber bands, masking tape or wires but for reasons which will be explained later, it is often advisable not to make a permanent fixture of the sacrum at this stage.

Now the last lumbar vertebra is taken and placed on the mount so that its post-zygapophyses articulate with the pre-zygapophyses of the sacrum and there is a small space between the centra representing the intervertebral disc. This vertebra is held in position by one of the methods suggested for positioning the sacrum. Its permanent fitting will consist of two clips made of the thinnest and narrowest metal which will support the bone.

The clips are attached to the main support by small screws. There are two possible points of attachment. If the main support is 5 mm or more in thickness the clips may be bolted to the sides by drilling and tapping holes at the correct places; when the metal is too thin for this the clips are attached by screwing them to the underside of the main bar, if there is a choice the first method is to be preferred, although the first stage in making the clips is not easy for a beginner. The length of metal required is determined by placing the end of a tape measure at the middle of the centrum at a point which is just above half its circumference and, pressing the tape against the bone, carrying it round until it meets the main support. Note this measurement and add to it 3–5 cm, according to the size of the bone; the measurement should be made on both sides of the vertebra. Cut two pieces of the appropriate sized metal to these lengths and on one of the broad faces of each piece mark off a length equal to that required to contact the bone surface plus

Mounting a quadruped 177

about 0.5 cm. Heat the metal to redness at this point and using two pairs of pliers, bend it sideways to form an L shape (Fig. 19). This is not easy since the metal tends to buckle at the bend and has to be flattened by re-heating and hammering on an anvil. In the short arm of the L, two holes are drilled. They should be at least 1 cm apart and should be wide enough to give clearance for the size of screw chosen for the fitting. The BA range of screw sizes is the best for this type of work, numbers 6 to 0 are those most commonly required.

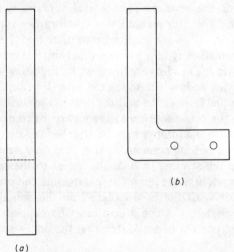

Fig. 19. First stage in making a supporting clip. (a) Metal marked for bending; (b) Metal bent to basic 'L' shape and drilled.

At a point on the long arm of the L, above its junction with the short arm, a distance equal to the thickness of the spinal supporting bar is marked. The long arm is bent forward at this point so that the clip may be placed with the short arm flat against the side of the bar and below the middle of the centrum of the vertebra. The position of the end of the short arm is marked on the side of the main support, and then the long arm is bent, using two pairs of pliers, until it fits the body of the vertebra. The fit is repeatedly checked as the bending proceeds by placing the short arm of the clip tightly against the side of the main support with its end upon the mark representing its correct position. The process is repeated on the other side of the vertebra. When a satisfactory fit has been achieved on both sides, the clips are held in their correct positions and the location of the screw holes marked on the side of the spinal support. The outline of the metal may be lightly marked in pencil on the bone.

The bones are removed and the bar unclamped. The marks on the sides, indicating the position of the screws, are centre punched and drilled with holes which are the tapping size of the screws which are to enter them, and these holes are tapped, forming the threads for the screws. The central support is replaced in its clamps and the sacrum and the vertebra are returned to their positions; the clips are screwed into place and the fit against the bone examined. Any faults are rectified and if there is excess metal it is cut away. The clips are finished, by rounding and smoothing, with a file the ends which touch the bone (Fig. 20). The vertebra is finally clipped in position and the next in the series is mounted by repeating the process. Alternatively two or three of the vertebrae in front of the last lumbar may be attached together with an easily resoluble gap-filling cement, so that it is only necessary to make one other set of clips to hold the series to the mount. When a group of vertebrae are joined in this manner the process should be carried out with the bones placed on the spinal support. The posterior vertebra of the group must be clipped to the support and it is important to make sure that the bones are evenly spaced

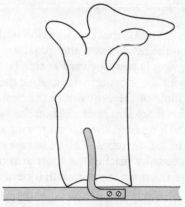

Fig. 20. Vertebra attached to spinal support by clips.

and that their neural spines are upright and in line. The jute-floc based cements are excellent for this purpose. Using the methods described, the entire vertebral column is mounted.

Stage 2. Establishing the Height of the Vertebral Column from the Ground and the Fitting of the Main Supports

At this stage a board, which is slightly larger in all dimensions than that required for the finished mount is needed and this is covered with thin cardboard attached with drawing pins. The card may be made up of several separate pieces joined with masking tape. A centre line is drawn along its length, and the spinal support bar is placed above this at a height which is estimated to be approximately that at which the limbs, when in position, will place it. Small mounts may be held with retort stands or laboratory scaffolding, larger ones are better suspended by cords or wires from an overhead beam; which may be made of slotted angle iron and its end supports of the same material. Whichever method is used it must be arranged that the spinal support may be raised or lowered easily. The only bones that need to be in position at this stage are the first five thoracic vertebrae and the sacrum.

It will be remembered that in stage 1 it was suggested that no holding clips should be made for the sacrum at that point in time. This is because the type and position of clips which are possible in this area depend upon the structure of the pelvis. In mammals the three elements of the girdle, the ilium, ischium and pubis are fused together; the ilia articulate with the sacrum and, below, the two sides of the girdle meet in the pubic symphysis. This makes it possible, when mounting a small mammal, to cement the sacroiliac joints and the pubic symphysis, and to treat the assembly as one entity for this purpose. With the sacrum cemented to the ilia, the areas in which it is possible to place clips are easily seen and they may be made accordingly. To achieve rigidity in the pelvic area it may be necessary to fit light metal bars on the insides of the ilia and to attach these, either to the spinal support or the rear upright which will be described below.

The ilia are fixed to the light bars by clips fashioned as previously described. In reptiles the pelvic elements are not fused and it is always necessary to make special supports for them, but before the fixing clips for the sacrum are made the joint between the sacral ribs and the ilium should be made temporarily in order to establish the points at which clips may be placed without this being fouled.

The sacrum and the pelvic girdle are assembled and attached to the mount, and the first five thoracic vertebrae are clipped into position. The limbs are assembled on temporary supports as previously described and in postures determined by the pose. The height of the spinal support is adjusted so that the heads of the femura will articulate with the acetabular facets and the scapulae are at their correct height in relation to the thoracic vertebrae. The limbs are then removed and it becomes possible to fix the spinal support to two uprights which will hold it at this height.

The uprights are placed so that one is below the sacrum or in some cases slightly in front of it, and the other below the junction of either the first and second or the third and fourth thoracic vertebrae. The rear standard should be at least 3 cm away from the front of the pubic symphysis.

Because a tube is more rigid than a rod of the same diameter, the upright supports for most mounted skeletons are made from gas barrel. As well as being attached to the spinal support these tubes have to be fixed firmly to the base, and this is made easy if one end of the gas barrel is fitted into the type of round metal flange used in gas fitting. Before the flanges are attached they should be drilled with three or four evenly spaced counter sunk holes which take suitable sized wood screws (Fig. 21).

The gas barrel may be attached to the flanges by cutting a thread at one end with gas fitter's dies; the flange is sold already threaded and is screwed onto the tube. The two components may be joined by welding or brazing if dies of this type are not available; whichever method is used the tube must be at right angles to the base at all points on its circumference.

Two of these assemblies are made so that the upright tubes are about 2.5 cm longer than is required to meet the spinal support at the selected places. These points are marked on the side of the support and the bones are then removed. The upright tubes are placed to one side of the bar with their centres in line with the marks. The lines along which the bottom of the bar touches the uprights are marked on the latter and the excess metal cut off

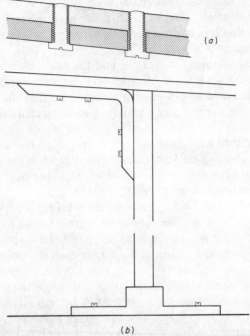

Fig. 21. (a) Section through upper arm of bracket and spinal support bar; (b) Method for the attachment by brackets of the spinal support to the uprights and the latter to the base board by gas flanges.

through the marks. The tubes should now fit under the bar and the tops should be in contact with it at all points; the centre line, which has drawn on the board should pass through a diameter of both flanges and the points on the back and front of the flanges which coincide with this line should be marked with a scriber. Having checked that the uprights are in line they may now be screwed to the board.

The spinal support must now be attached to the two uprights by metal brackets; which are made from the same sized metal as that used for the spinal support or some which is not quite so wide. The lengths required obviously depend on the size of the mount, but there should be sufficient to make each arm of the bracket long enough to take two widely spaced screws, and it is an advantage if the arm which is to fit the mount in the region of the neck is made long enough to lie beneath the fifth, sixth and seventh cervical vertebrae. This increases the rigidity of the mount in this area and the neck is less likely to sag from the weight of the skull.

The required lengths are cut and the portions which are to fit the uprights are marked off from one end and centre lines drawn down the whole length of each piece. In each of the areas which are to fit the uprights, two widely spaced holes are drilled on the centre line. The diameter of the holes should be the recommended clearance size for screws to be used. The bars are bent at the lines dividing the two sections, so that when the pieces which fit the uprights are placed in position the only part of the upper arm of the bracket which touches the spinal support is the origin of the bend. The backs of the sections through which the holes have been drilled are then hollowed slightly with a round file. The fitting for the front standard is taken and placed upon its forward face and pushed up until it lightly touches the underside of the spinal support and having made sure that the sides of this arm of the bracket are vertical, its outline is marked on the gas barrel and the position for the screws indicated by running a pencil or a scriber through the holes. The upright is removed from the base board, centre punched and drilled at the points marked for the screws. The drill used must be the tapping size for the selected screw. The holes are now tapped and the upright returned to its position on the board. The arm of the bracket which is to fit below the neck is now gradually bent until it fits the spinal support and at each stage of the bending the fit is checked by placing the bracket onto the upright so that the lower arm fits the outline drawn on the tube; in the last stage it should be screwed into place to check the accuracy of the fit. When the fit is as perfect as it can be made, two widely spaced clearance holes are drilled in the arm and countersunk on the outer side to take the appropriate screws; the ends of the bracket are chamfered to neaten their appearance and the fitting screwed back into position. Pencil marks drawn through the newly drilled holes mark the drilling points for the screws on the underside of the spinal support. This is removed from its clamps or cords and holes drilled and tapped at the marked points. It is then returned to its position and screwed to the front bracket (Fig. 21). The bracket which makes the connection in the sacral region is made in the same way. It may be attached according to convenience, either to the back or the front of the upright and since it may be hidden by the pelvic bones, there is often no need to countersink any holes; cheese-headed screws may be used at all points. When this bracket has been attached to the spinal support the mount is at the correct height above the base and all auxilliary supporting devices, clamps, overhead beams, etc., can be removed.

Stage 3. Attaching the Skull

It will be remembered that in stage 1 a length of metal was allowed for the support of the skull and that a mark was made on the spinal bar indicating the position of the occipital condyles. The bar is now bent at this point so that the condyles articulate with the atlas vertebra and the skull is at the required angle. Depending on the thickness of the metal and the availability of assistants, this bend may be made either while the bar is still attached to the upright or it may have to be removed; in any case there should be no bones on the bar and any clips in the near vicinity should be removed. Should it be possible to work with the bar on the uprights an adjustable spanner is fastened just behind the marked point and is held by one person. The projecting piece of bar is inserted into a strong metal tube which is longer than the bar and the end of the tube is pushed down the bar until it is just in front of the mark. The spanner is held firmly and a second person pulls the tube downwards until the bend is at the correct angle. If the spinal support must be removed from the uprights all clips should be numbered in pairs and removed to avoid their being distorted by accident; the bending is then done by clamping the bar in a vice at just below the mark and using the previously mentioned tube as a lever.

The bar is then returned to the uprights and the articulation of the atlas and the condyles and the angle of the skull to the horizontal is checked. If these are correct, the bar is removed from the uprights and any excess metal is cut from the end. It is now formed to fit the underside of the skull and the palate, as described in the section dealing with skull mounts. The bar is obviously unwieldy so two people or an elaborate arrangement of supporting clamps are required if this operation is to be carried out successfully. The supporting bars and retaining clips are made in the same way as described for an isolated skull, but more attention must be paid to the planning and design of the clips. The skull must be locked firmly to the mount by them and the mandibular rami must be firmly held. The latter are often attached to the skull by cementing a steel bar between the two rami in the area of the ascending processes; a short bar may also be placed towards the front of the jaws, and these bars are then suspended on metal hooks which are attached to the main supporting bar with screws. Two hooks are required for the rear bar and one for the front. The other methods previously mentioned are also valid.

There is a limit to the size of mount to which this technique of skull mounting can be applied. It has been done successfully with skeletons up to the size of a small horse, but larger and heavier specimens require that the skull mount be made separately and then attached to the main support by a joint which may be fixed either by screwing or welding. In extreme cases a third upright may be needed to support a skull, but this is very ugly and should be resorted to only when every other method has been tried.

When all work on the skull has been completed the spinal bar is returned to the uprights and screwed in place and the vertebrae are placed in position.

Stage 4. Mounting the Ribs

This is the most difficult and tedious of all the stages in the mounting of a skeleton, for as far as I am aware, there is no efficient method of temporarily supporting the whole of the rib cage, and this means that each pair of ribs has to be correctly positioned before the next can be dealt with. The normal support for a rib cage consists of either one or two bars of light gauge narrow metal placed along each side of the mount and connected to the main support by two bars of the same metal, one positioned below the first rib and the other below the last. Since it should be clear from what has been said in the preceding stages that the vertebrae have to be removed from the spinal support and this itself detached from the uprights every time a hole has to bored in it, these facts will not be repeatedly mentioned in the following section, but it will be assumed that the reader is able to deduce for himself when this should be done.

The points on the spinal support which coincide with the position of the facets on the vertebral centra into which the heads of the first and last ribs articulate, are marked. At these marks, two pieces of metal bar are attached, which are long enough when bent, to fit the contours of the bones on the undersides to extend to about three quarters of their lengths. About 2.5 cm from the ends, holes large enough to take a small BA screw are bored. Two lengths of stout wire or strips of metal which will bend easily (aluminium about 1 cm wide and 1–2 mm thick is ideal) are taken and bent to form an even curve which reaches from the holes in the front supports to those in the back and bulges outwards so that, when the rib in the centre of the thorax is placed in position, it touches the bar at about the middle of the shaft of the bone. The piece of metal has the appearance of a sabre blade bent sideways. The length of metal required to form the curve on each side of the mount is ascertained by applying a piece of tape to the underside of the pattern and measuring its length. This length of mild steel plus 3 cm is cut for each side of the mount and, at a point equal to half the width of the supports for the first rib from the end of the strip, holes equal in diameter to those in the ends of the rib supports are drilled. The steel is now bent to fit the wire or aluminium pattern and is fixed by a nut and bolt to the front support and to that at the back by a small screw clamp.

The heads of the second pair of ribs are placed in the articular facets. If the curve at this point is correct they will articulate properly, part of the shafts will rest on the bar and they will be symmetrical to the vertebral column. Should this not be the case the curve is adjusted until these conditions are fulfilled. For this to be done the bar is released from the rear support and after the alteration has been made, is clamped back into position. This process is continued with each pair of ribs until the whole cage is completed.

The position of the holes in the rear supports is marked in the linear centre of the curved bars. Holes are drilled at these points, the excess length of metal cut away, and the bars fixed with nuts and bolts.

It has been said that fossil bone should not be drilled or cut, but an exception is sometimes made in the case of ribs, and small holes, into which wires can be fitted, are drilled into the heads and into the articular facets of the vertebrae. When this may be done it avoids the making of a second curved bar to fit just below the heads of the ribs and the tops of the bones can be held in position by pegging them to the vertebrae with wires; this leaves only the shaft to be secured to the curved bar. Normally this is done by placing each rib, from first to last, into its correct position and scribing lines on the bar along both sides of each bone. Small holes are bored through the centres of these lines, through which wires are passed so that they cross the rib, and are then twisted together behind the support thus securing the bone in position. Cotton covered copper wire is best for this purpose because it is easy to camouflage by painting.

Stage 5. Mounting the Limbs

Now that the parts of the skeleton representing the trunk are in the right position the limbs may be mounted, but before commencing this be sure that all work required to support the pelvic girdle has been completed. The four limbs should now be set up, on temporary supports so that they adopt the required pose. The areas occupied by the manus and pes are marked on the base board and all the limbs but one back leg are removed. This is mounted in the manner described for mounting single limbs, care being taken to make a proper articulation between the head of the femur and the acetabulum. To keep the limb steady it may be necessary to make a connection between the iron which runs up the back of the leg and either the bracing within the pelvis or with the upright support. The other rear limb is then mounted.

When mounting the fore limbs it is important to place the scapulae at the correct height and angle in relation to the spinal column and to see that the articulation between the glenoid fossae and the heads of the humeri are in the right positions in relation to the first pair of ribs. These positions are largely determined by the pose of the skeleton. The mounting procedure is as previously described; extra connections to the main upright may be necessary to hold the scapulae in the correct positions.

Stage 6. Supporting the Caudal Vertebrae

Although the mounting of the tail is discussed at this point, it may be done at any stage after the sacrum has been finally attached to the mount. The method used depends upon the length of the tail. The caudal vertebrae decrease in size from the first to the last. The result of this is that, if the tail is long, at some point the vertebral centra will be narrower than the metal bar used to support the spinal column. The width of the bar may be progressively lessened by tapering with a file, but if the tail is very long the thickness of the bar will be disproportionate to the height of the last few vertebrae and the result will be unpleasing to the eye. This can be overcome by cutting off the spinal support at the point where the vertebra is just wide enough with respect to the bar, and attaching, at this point, a narrower bar of less thickness. The connection may be made by filing a flat recess on top of the main bar, of length about 3 cm and depth equal to the thickness of the smaller bar; the two pieces may be joined by counter sunk screws, riveting or welding. The lighter metal is then bent to the correct shape and the bones attached by small clips or by wires according to their size. For the smaller ones a variation on Sizer's method (see page 155) may be used; wires may be cemented to the undersides of the centra and these passed through holes bored in the bar.

Stage 7. Mounting the Sternum

Like the mounting of the tail this may be done at an earlier stage, but the rib cage must be in position before it can be placed accurately. The sternebrae are supported on two narrow strips of mild steel, one fitted to the front of the forward upright and one at the back. The height above the base is determined by the ends of the ribs and the estimated length of cartilage which would have connected them to the sternum in life. Because of the unavoidable presence of the forward upright in this position at least one of the sternebrae has to be cut to fit round it. To avoid damage to the bones, those effected should be cast in fibreglass resin and the casts used on the mount in place of the bones.

Variations in Procedure when the Spinal Support Bar is placed within the Neural Arch

The vertebrae are supported in plasticine as indicated in stage 1 above. A line is drawn on the surface on which they stand by running a pencil or a piece of chalk along the undersides of the centra. A metal bar is selected which will pass through the narrowest neural arch in the series, from cervicals to lumbars; this is now cut to a length which allows for the length of the skull, and is bent to fit the line mentioned above.

The vertebrae are threaded onto it starting with the last lumbar and working forwards. If the arch above the sacrum is wide enough to allow the passage of the bar this too may be threaded on but, in most mammals, this is not likely to be possible; a clamp is therefore placed at the point where the last lumbar contacts the sacrum. This prevents the lumbar vertebra from moving too far back on the bar. The assembly is now held in a horizontal position with clamps and the vertebrae adjusted so that the zygapophyses articulate and the centra are evenly spaced.

The spacing may be maintained by sticking discs of cork to the posterior aspect of each vertebra. It may be necessary to adjust the level of some of the bones and this can be done by sticking small pieces of wood inside the neural arches; the same device may be used to keep the neural spines upright.

The point on the bar coinciding with the posterior end of the last lumbar vertebra is marked and the bones removed. The metal is bent at this point to an angle which will place the downward section along the posterior face of the centrum of the last lumbar vertebra when this is placed in position. At a point slightly below the body of the vertebra it is bent upwards so that it passes under the sacrum and places this bone in its right relationship to the last lumbar and at about 3 cm from the posterior end of the sacrum the metal is cut away. Usually the thickness of the bar between the last lumbar vertebra and the sacrum is greater than that of the intervertebral disc in life. To narrow this gap the centrum of one or both of the vertebra would have to be recessed, therefore they should be cast and the casts used in the mount.

The assembled vertebrae on the mount are now suspended, as in stage 2 above, and the height of the upright supports established in the same way. Should the width and position of the bracket connecting the spinal bar to the front upright space too widely the two vertebrae between which it must pass, the bones must be cast and the casts recessed to correct the fault. The rest of the procedure is as described previously, with the one exception that in some cases, when the foramen magnum is open and the brain cavity empty, the skull may be mounted by threading it onto the bar projecting from the neck. To avoid damage to the bone a piece of plastic tubing should be pushed over the bar and about 3 cm allowed to extend beyond the end of the metal. The lower jaw is fitted by suspending it, as previously described, from a lighter piece of metal which can be attached to the main support at a point between the condyles of the skull and the atlas vertebra.

The sole advantage of this method is that it avoids the making of clips for the vertebrae and no spinal support is visible. Its disadvantages are that the angle of the neck and skull are limited because the vertebrae cannot be threaded past a very sharp bend; at least six bones have to be cast to avoid cutting the originals and the vertebral column has to be completely dismounted if any vertebra is required for study.

The real purpose of the card covering the base board may now be explained; it is used as a template when the mount is transferred to its final position.

Mounting Bipeds

The general procedure is as for quadrupeds, but the vertebral column is raised at the front to place it at a reasonable angle to the base and the pectoral girdle and forelimbs are mounted on supports which are attached to the forward upright, which is positioned with this fact in mind.

Besson (1963) published an account of the use of fibreglass in mounting fossil skeletons. This method is worthy of consideration but careful design is needed or the support may become rather more attractive than the specimen.

Fossils preserved on slabs of rock are often to be seen mounted in blocks of plaster of Paris. This is sometimes done to add strength to the specimen and sometimes to make the surrounding material rectangular. The method is simple and the results are pleasing to the eye. Because plaster is brittle and the edges of these blocks are easily chipped they are often placed in wooden frames. It is possible to cast the plaster block inside the final frame but if this is done the wood must first be thoroughly painted with a waterpoof paint; polyurethane or epoxy varnishes are the best for this purpose. Although it takes longer it is better to cast the block and fit a frame round it afterwards.

A smooth flat surface is required on which to make the mount; plate glass is the best, but a flat board covered with a plastic material such as is used for making the tops of kitchen fittings will do. The back and sides of the specimen are treated with a heavy coat of a water-resistent varnish, which is easily resoluble; shellac does very well or one of the 'Butvars' may be used. The surface of the glass or plastic sheet is lubricated with olive oil and the specimen placed, display side up, on several pillars of plasticine which raise it to a height equal to the desired thickness of the finished mount; the top of the specimen must be level. If needed to give added strength, metal bars, painted with waterproof paint, may be placed slightly below the block and held above the casting bed on pillars of plasticine.

A wooden frame, the depth of which equals the required height of the finished block, is now made so that it may be dismantled easily; this can be done if the sides are held together by screws. The frame, after the sides have been oiled very thoroughly, is put round the specimen and held in place by weights at the corners. The areas where the bottom of the frame touches the casting bed are sealed with plasticine on the outside; the outsides of the corners of the frame where the sides meet are treated in the same way. The face of the specimen is now protected by a piece of tissue paper stuck to its edges with masking tape. Plaster of Paris is mixed and poured into the frame and if more is required the next mix should be poured before the first has set.

The block is left to set for several hours and then the frame is removed. The whole process is illustrated by Fig. 22. The mount should now be moved slightly by carefully pushing on the sides; if it will not move it sometimes helps if water is run round the base. It is then tipped up onto one of the longer sides and the plasticine pulled out of the holes in the back. The holes are filled with a stiff mixture of plaster of Paris which can be picked up on a spatula.

The block is then lowered onto its back on a series of wooden battens. There should be several of these placed along the length because plaster which has not fully dried will distort under its own weight if it is not properly supported. The plaster is now left to dry;

Mounting bipeds 191

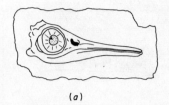
(a)

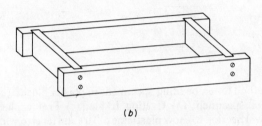
(b)

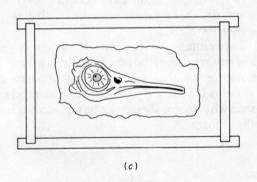
(c)

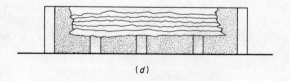
(d)

Mounting Fossils for Exhibition

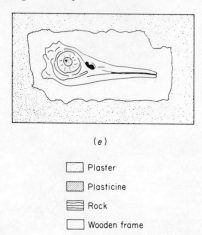

(e)

▢ Plaster
▨ Plasticine
▤ Rock
▢ Wooden frame

Fig. 22. Method for embedding specimens in plaster blocks. (*a*) Original condition of specimen; (*b*) Casting frame; (*c*) Frame placed round specimen; (*d*) Section to show plasticine pillars and level to which plaster is poured; (*e*) Finished specimen.

this will take several days if it cannot be done in an oven set at about 50°C. When it is dry the surface is smoothed with glass paper and painted to match the matrix.

The big disadvantage of this type of mount is that, if the specimen is of any great size, too much weight is added. Therefore there is room for another method using modern materials. Although, as far as is known it has not been done, there seems no obvious reason why plaster should not be replaced, at least in part by polyurethane foam.

References

Besson, L. C. (1963) 'A technique for mounting skeletons in fibreglass', *Curator*, VI, 231–9.

Claringbull, G. F. (1948) 'Perspex supports for exhibition', *Mus. Journ.*, **48,** 8–10.

Sizer, C. A. (1960) 'Mounting geological specimens', *Curator*, III, 371–9.

9
Casting

Casts of fossils are made for many reasons. They can be used to make unique specimens available for study and exhibition in many places, and are thus valuable for exchange purposes between one institution and another; their use as teaching aids is of great importance and avoids putting the rarer specimens at the inevitable risk which exists when they are passed from hand to hand, and as has been said in the section on mounting fossils, they may be used in place of bones in a mounted skeleton. In most museums there are specimens which are in constant demand for study by scholars and at the same time they are of interest to the general public and should be on exhibition. The repeated removal of a specimen from exhibition for study is not desirable, because it increases the possibility that it may be damaged and while it is out of the case, the public not only cannot see it but when it is part of a story told by an exhibit, the whole effect is spoiled. There are very strong arguments for using casts in exhibitions in these instances.

To fulfil any of these purposes the cast must be a faithful reproduction of the original and when it is to be exhibited, it must be coloured to as near an exact match to the original as is humanly possible. Bad casts are worse than useless for they can be misleading.

All casts have to be taken from moulds and nowadays there are many excellent materials from which these can be made, but up until the late 1940s there were only two in common use, these being plaster of Paris and gelatine in some form. There are occasionally references to the use of gutta percha.

Plaster Moulds

Plaster of Paris is a rigid material and cannot be pulled out of undercuts so for all but the simplest shapes, elaborate piece moulds had to be made. This required great skill and was time-consuming, and one has to admire the enormous ability and patience of the men who did this work. The standard of the resulting casts, however, was limited by the properties of the material because, to ensure separation of the cast from the mould, the latter had to be shellaced until a polished surface was formed. Even when this was done by careful workers, it inevitably resulted in the obscuring of any fine ornament on the fossil. This did not matter when large vertebrate bones were involved, but the identification of many fossils now depends on such detail. In spite of its limitations and the time-consuming nature of the technique, it would be a great pity if the art of piece-moulding fossils were allowed to die out and it is to be hoped that the few remaining experts (among whom the present writer is not numbered) will be willing and encouraged to pass on their skill and knowledge to the young.

Gelatine Moulds

Sheets of solid gelatine, if soaked in cold water for some hours, will absorb some of the liquid become soft and may be torn into small pieces easily. If the excess water is shaken from the material and it is placed in a double saucepan, the bottom half of which contains water, and heated, it will melt and become pourable. On cooling it forms an elastic mass which can be pulled out of undercuts. Gelatine is available in differing degrees of refinement from crude glues to an edible type and there are many different opinions as to which grade is best for casting. Suppliers of artist's materials sell what they call 'Moulder's' gelatine and this is quite as good, if not better than the edible variety which is generally very much more expensive.

There are a number of substances which can be added to gelatine which are intended to improve its performance as a moulding medium. Glycerol, sorbitol and laundry soap are included in some formulae (Clarke, 1938). The object of the additions appears to be that they reduce the speed at which the gelatine loses water and shrinks but for most purposes gelatine soaked in water works as well as more elaborate mixtures. It is, however, necessary to add a substance which will inhibit the growth of moulds; clove oil is the most pleasant as far as smell is concerned, or a few drops of a solution of phenol in water may be used.

Gelatine is a form of glue, so if it were poured straight onto a specimen and allowed to cool it would stick to the surface; a separator is therefore necessary. This may be almost any type of oil or grease which will not stain the fossil; olive oil is most commonly used, silicone fluids are more efficient.

Plaster of Paris is the traditional material for making casts from gelatine moulds. The plaster is mixed with water and since heat is generated as it sets, although the gelatine is coated with oil before the plaster is poured, both the water and the heat tend to blunt very sharp detail, thus making it seldom possible to get more than one or two casts which are up to the standard required by scientists from a gelatine mould. Efforts have been made to reduce the effects of heat and water by rendering the surface of the gelatine less soluble and less inclined to melt, for example, by treating the moulds with solutions of alum, potassium dichromate or other chemicals which are used in tanning leather. The results are not good, for unless every trace of the chemical which has not reacted with the gelatine is removed, the surface of the plaster when set, is very weak and ornamentation is lost.

It is possible to make small casts from gelatine moulds in cold-setting resins. The size is limited because the larger the volume of these resins the greater is the heat produced on setting; the range can be extended if fibreglass casts are made. The epoxy resins are better than the polyesters in this context as they are not inhibited by the presence of a small amount of water but if either epoxy or polyester resins are used, phenol should not be added to the gelatine.

As a moulding material gelatine has many attractive qualities. It is re-usable and many moulds can be made from one batch; the pouring temperature can be as low as 40°C, which means it may be used with safety on nearly all fossils, and it takes an excellent impression of fine detail; additionally it is clean and pleasant to use. Offset against this is the difficulty of getting the detail from the mould to the cast, and for this reason it is far less popular now than it was, as there are several modern materials from which many excellent casts can be taken without any trouble.

Modern moulding materials may be divided into two physical types; those which are mixed and used cold and those which require heat, either to melt them or to convert them from a pourable paste to an elastic solid.

Water Mixed Materials

There are some compounds, presumably based on sodium alginate, which when mixed with water form a gel. They are made for use in dentistry and there are variations on them which are sold, in some countries, by manufacturers of materials for artists. These compounds are powders and a measure both for the solid and the required amount of water is normally sold with them. The flexible solid formed on setting is not very strong and will not pull out of a deeply recessed undercut but the impression taken from the fossil is good and casts taken from them show fine detail. The moulds have to be kept wet or they will shrink. On some fossils, no separator such as olive oil is required, but it is always wise to try any moulding medium for which this claim is made on some unimportant material first. One trouble encountered with some of these media is that the moulds tend to have air bubbles in them. This tendency can be decreased if care is taken in the mixing and, after mixing, the container is banged sharply down onto the bench top and the bubbles which are thereby forced to the top of the mixture are skimmed off before it is poured.

These compounds are much favoured by travelling palaeontologists who wish to make casts of specimens in the institutions they visit; in particular, scientists studying teeth find these substances very useful as no apparatus apart from the measure provided, a vessel for mixing, a stirring rod and some plasticine is required. Plaster of Paris is used for the casts from these moulds.

Brand names of two of these compounds are 'Zeelex' and 'Duplit'. There may be others and enquiries should be made about them to dental suppliers and firms who market materials for model makers and sculptors. Every brand has an instruction leaflet packed with it and this should be read carefully.

Rubber Latex

This has long been used for casting fossils and several papers have been written on the techniques employed (Fischer, 1939; Fuehrer, 1939; Quinn, 1940; Quinn, 1952). Its great advantage is its unrivalled elasticity and its great disadvantage is that there is always considerable shrinkage as it sets. Workers often say that this or that brand of latex does not shrink on setting, but all those that the author has been able to test do not live up to this claim. It is said that if the latex is built up by painting thin layers on the specimen, each being allowed to dry before the next is applied, and pre-shrunk cotton fabric is incorporated between the layers there is little or no shrinkage. This claim is supported by a number of excellent casts of bone implements made by Mr Martin Burgess at University College London, several years ago.

The surface of a fossil on which latex is to be used must be sound and may be consolidated with Butvar B 98 or any other consolidant which is not water soluble. The application of 5 per cent (v/v) ammonium hydroxide before the rubber helps penetration into small detail and reduces the likelihood of bubbles forming in the mould. Normal separators like mineral or vegetable oils, cannot be used since they attack rubber; on a sound surface nothing but the consolidating plastic is needed but with care silicone fluid may be used. It is applied with a brush which has been dipped in the fluid and then wiped almost dry, or as a weak solution in chloroform. The silicones are extremely good water repellents, and if too much is used it is difficult to apply the first layer of rubber with a brush, but when the fossil is flat and the first layer can be poured there is no problem.

The standard method of applying latex is to brush it on layer by layer. The brush must be washed out immediately in a weak solution of ammonia. Bandage or surgical gauze may be applied between the layers to give support to thin walled moulds; an alternative method is to paint on two layers, the first being dry before the second is added, and then to back this with a mixture of jute floc and latex. The mixing of these two substances is made easier if the floc is first moistened with a 5 per cent solution of ammonia. Drying at all stages may be accelerated by using a current of hot air from a hair dryer. The rubber may be poured on instead of painted and layers up to 6 mm thick will dry fairly quickly. The speed of drying obviously depends on the temperature and humidity of the surroundings and should be established by experiment.

The translucency of ordinary latex is sometimes a disadvantage and this is easily overcome by adding powder colours to the liquid rubber. Titanium dioxide will make the hardened rubber white and burnt umber produces a chocolate brown which is excellent for photography. The powders are first mixed with 5 per cent ammonia solution and then strained through a piece of nylon stocking into the latex and stirred in. Some makers of rubber latex sell filler pastes and coloured latexes. It is not advisable to mix any form of black into rubber as it has been found that moulds and casts of this shade decay very quickly when exposed to light. Latex may be thickened, as described in Chapter 3, by adding aerogel silica.

When the type of mould required necessitates the partial burial of the fossil in a soft modelling material or the use of a retaining dam, plasticine cannot be used unless the parts of it which come in contact with the latex are first coated with shellac or a silicone grease; this is because its oil content will attack the rubber and the damage does not occur until some weeks after the mould has been made. For practical reasons rubber moulds are usually thin-walled and are easily distorted so they have to be supported by a plaster 'Mother' or mould box before casts can be taken from them.

Plaster of Paris casts may be taken from rubber moulds without the use of a separator. Rubber is not really suitable for moulds from which fibreglass or solid resin casts are to be taken, although this can be done if the insides are first coated with a solution of polyvinyl alcohol and this allowed to dry, but even then there is the risk of damaging the mould if the film is broken. Set rubber latex will expand if it is soaked in paraffin oil. This property can be used to make enlarged casts by expanding the mould, taking a cast and from the cast making another mould which in its turn is expanded. This can be done several times without loss of detail.

Polysulphide Rubbers

The polysulphide rubbers are available in several colours and viscosities. Some of the colours will stain a specimen, however the 'Smooth-On' range includes one which is specially made to avoid this trouble. All types have two ingredients which must be mixed in the correct proportions for the best results; the makers supply technical details either with the product or on demand. The polysulphides have been used by sculptors to cast quite large pieces. The moulds can be used to produce casts in plaster, resins or fibreglass. Some care is required in the mixing and pouring of these rubbers to avoid the formation of air bubbles in the moulds.

Dental Rubber Impression Compounds

There are some dental compounds which are packed in two tubes, which, when one is mixed with the other, form a mass with a texture rather like chewing gum. Water is both an accelerator and a separator for them. The colour is usually brown. These compounds are excellent for taking impressions from small natural moulds or for making one piece moulds of something like the crown of a tooth. They are very easy and quick to use, but are difficult to clean off the surface on which they have been mixed. A cheap throw-away mixing board may be made by cutting a piece of hardboard to fit inside a suitable sized polythene bag.

Before the mixture is applied to the fossil, it must be in such a condition that the surface is not tacky when pressed onto a piece of wet glass or glazed porcelain. This action also produces a perfectly flat and unmarked face of the plastic mass that is then pressed onto the specimen which has previously been thoroughly wetted with water. Information as to the availability of these materials is best sought from dental suppliers. Some of these compounds have a limited shelf life and since they are fairly expensive, it is best to buy only when they are needed. They are very useful to the travelling palaeontologist.

Silicone Cold Cure Silastomers

These substances are without doubt the most easily useable and the most efficient of moulding materials for most fossils. It is almost impossible to make a bad mould with them but they have two disadvantages. One is the cost and the other they have not sufficient mechanical strength to allow them to be pulled out of deeply recessed undercuts; the second of these can be an advantage as well as a disadvantage for the silastomer will always break before the specimen. They are available in three viscosities. Dow-Corning cold-cure Silastomer 9161 used with Catalyst N.9126 is an excellent example and is widely used.

The uncured silastomer is a pourable opaque liquid which is activated by the addition of from 1–4 per cent of the liquid catalyst. The silastomer must be weighed, but it can be taken that one millilitre of the catalyst is near enough to one gram for all practical purposes. The cured material is very difficult to clean out of glass vessels and a great deal of time can be saved if the weighing and mixing is done in disposable paper cups. The catalyst for the same reason should be stirred in with a wooden spatula; iced lolly sticks or tongue depressors are ideal. Unfortunately, the measuring cylinder used for the catalyst must be looked upon as having only a limited useful life, and is often broken before the end of this is reached.

The setting time of the silastomers is affected by the temperature and the amount of catalyst used. Not only does the catalyst control the setting time but it also alters the amount of shrinkage. The smaller the amount of the catalyst the smaller the shrinkage, which is not very great even if the full 4 per cent is used, but 2 per cent not only reduces this but extends the pot life and so the time available for application to the specimen. Up to 10 per cent of silicone fluid may be added to obtain a less rigid, but not a stronger mould. All mixing must be thorough, especially when lower quantities of catalyst are used.

Silicone cold-cured silastomers may be poured onto the specimen to make a mould. If enough is used to form a mould with walls about a 1.25 cm thick, the material is sufficiently rigid to make support by a plaster mould box unnecessary. Owing to the price of the material this may at first sight appear extravagant, but the man hours saved offset the cost at least where specimens of up to 7.5 cm square are concerned. For larger specimens the silastomer may be brushed over and backed by adding further layers which incorporate surgical gauze. The silastomer will weld to itself when the uncured material is applied to some that has already set, so large moulds can be made in this manner, and because of the strength of the gauze, this type of mould may be pulled out of undercuts without damage to it or the specimen. These of course have to be supported by a 'Mother' of plaster of Paris which is built upon them before the specimen is removed.

A mould of a plesiosaur skeleton approximately 1.5 metres long has been made this way, and the time saved in making the mould more than offset the price of the silastomer as none was wasted in waiting for the moulding material to set. It was applied to the specimen in 100 g-batches and each had set before the next had been weighed out and mixed.

Water will act as a separator for the silicone rubber, which makes their use possible in moulding water-logged objects; for most dry specimens it is best to use a liquid detergent such as 'Teepol' which is brushed onto the fossil and allowed to dry before the mould is made. As has been already said, the silicone rubber will weld to itself if some uncured silastomer is applied to some that has already set. This fact may be used to repair a torn mould, but when a two-piece mould is required, the two halves must not be allowed to weld together. A solution of soft soap painted onto the surface of the set rubber which is to come into contact with that in the liquid stage will stop this happening. The silastomers are not affected by the oil in plasticine and this may be used to set up a fossil for moulding or as a containing dam round a natural mould. Clean separation between the two substances can be ensured by coating the plasticine with a solution of soft soap.

Silicone rubber will expand if it is soaked either in toluene or in paraffin oil (kerosene); the latter is used for preference as it is less volatile and the expansion lasts several days. A technique for making enlarged casts of fossils for exhibition based on this fact is described at the end of this chapter.

Casts from silastomer moulds may be made with plaster of Paris, solid resins or fibreglass. Although the material is inflammable it is not affected by low heat in the solid state. The makers do not, as a rule, sell silicone silastomers in small quantities, but they are available from some chemical suppliers and dealers in sculptors' materials.

Moulding Compounds requiring Heat

These fall into two classes, the hot-melt compounds which are solids which liquefy on heating, and on cooling revert to solids and pastes which solidify on heating and cannot be re-melted.

Hot-melt Compounds

The hot-melt compounds are manufactured in several grades. They vary in colour, hardness and melting point and generally they may be mixed to produce a substance with properties which are between the grades; they are all based on polyvinyl chloride. The lowest melting point obtainable is about 120°C, which is too high for safe use on some fossils. Only those that are well mineralised will stand up to this heat; so it would be unwise to use these compounds on most Pleistocene vertebrate fossils, and they should never be used on ivory or teeth. These substances have a sharp cooling point and if they are poured onto a cold specimen the detail taken by the mould will be blunted in places. If the fossil is small enough and will stand such treatment, a period of pre-heating in an electrically-controlled oven will overcome this difficulty. The makers supply a sealing oil and this helps to overcome the sharp cooling point trouble and reduces the tendency to the formation of bubbles on the mould surface. There is the need, however, to be very sure that this oil will not penetrate into the specimen to the extent that it cannot later be removed or will stain it. Treatment of the surface of the specimen with a solution of polyvinyl alcohol is to be recommended before moulding with a hot-melt compound. There is often no need for a separator as the compounds have a naturally oily surface; but in cases of doubt a separator should be used. Almost any oil or grease will do, but silicones are the best. For fibreglass resin casts, the inside of the mould should be dusted with talc.

Vinatex Ltd produce a series of hot-melt compounds under the name of 'Vinamould'. Two of them are particularly useful. One is coloured red and the other green; the green one is especially made for use with fibreglass. The makers also sell a handbook and market electrically-heated thermostatically controlled buckets for melting the material. Small quantities may be melted in a beaker in a heat-controlled oven, or in a double saucepan, the lower half of which is left empty or is partly filled with asbestos wool and is better heated on an electric hotplate rather than a gas-ring. Whatever the method of melting, it is necessary to stir the compound from time to time. When the compounds are hot they give off a rather unpleasant vapour and so for the sake of comfort they should not be used in badly ventilated rooms. The compounds are re-usable, and small quantities can be bought from suppliers of materials to sculptors. They are sold in blocks which have to be cut into small pieces before melting.

Plasticine is the best material for laying up fossils which are to be cast with hot-melt compounds. A plaster 'Mother' is required for support when casting from these moulds. Multi-piece moulds can be made with hot-melt compounds.

Casts may be taken in plaster, solid resin or fibreglass; the last has been used to cast very large sauropod bones from this type of mould.

Polyvinyl Chloride Pastes

These used to be available ready-made but some manufacturers have stopped this practice. Those which can be bought are rather better than those made in a laboratory, but they do have a limited shelf life. Stahl (1951) described a method for making such pastes and their use for moulding fossil fish. When buying material for making P.V.C. pastes it is important to obtain a very finely ground polyvinyl chloride powder, for the coarser types produce a mould which looks as though it was made from coagulated sand. Suitable grades of the powder are manufactured in Great Britain by Imperial Chemical Industries Ltd, Plastics Division.

The paste is made by mixing the powder with a liquid plasticiser, dibutylphthalate. The inclusion of 0.5–1.0 per cent calcium or aluminium stearate reduces the possibility of charring, but it may be omitted. The usual mixture has equal weights of the solid and the liquid. The mixing must be very thorough and the paste should be allowed to stand for at least 12 hours to allow air bubbles to disperse. These proportions make a paste having the consistency of condensed milk, and it looks as though it is too viscous to enter small detail, but this is far from the truth for it will enter the smallest cracks and reproduces the finest ornament. Instances have been known in which a label with a registered number written in indian ink has been faithfully reproduced with the number clearly legible. A less viscous paste may be obtained by reducing the proportions of the solid to the liquid; three parts by weight of the powder to seven parts by weight of dibutylphthalate is a practical mixture which results in a more flexible mould. For special purposes the solid content may be further reduced, but when this is done the cured mould must have completely cooled before any attempt is made to remove it from the specimen.

For full cure, that is conversion from a liquid to a solid state, the whole of the paste must reach a temperature of 150°C. This condition is indicated when a sharp point can be drawn across the surface and this does not break, also the unfilled pastes go translucent at this point. Solidification occurs at a much lower temperature but at this stage the material is not elastic and is mechanically weak.

Because of the high curing temperature, the paste is only suitable for moulding well mineralised fossils from the older rocks. As has been said in Chapter 3 it is exceptionally good for making casts from natural moulds. If a specimen needs consolidation this must be done with a substance whose melting point is above 150°C; a solution of polyvinyl alcohol in water may be used, but all the water must be allowed to evaporate before the mould is made. A modern form of carpenter's glue, such as 'Croid', may be used for repairs. All specimens should be checked to see that they have not been previously consolidated or repaired with a substance which will melt at the curing temperature.

Plasticine cannot conveniently be used for the partial embedding of specimens or the making of dams because, at the full curing temperature, it causes frothing and charring of the plastic but it can be used if the mould is first raised to only 90°C, at which point it forms a weak solid. The plasticine is removed and the plastic is taken to 150°C. Removal at this stage is not always easy and if even a small trace of plasticine remains the mould may be useless. Modelling clay can be used but this may result in bubbles due to the water content being formed in the plastic. Stahl used plaster of Paris dams, but these are time-consuming to make and, to a lesser extent, have the same disadvantages as modelling clay. A special putty may be made by mixing asbestos powder into silicone grease. The method of mixing is that used by pharmacists to make an ointment. Some grease is placed on a tile or a sheet of glass and some powder is placed beside it. A little powder and a little grease are mixed together with a steel spatula and the process is repeated, gradually increasing the quantity of each ingredient used until the mass approaches the consistency of putty. The mixing at this stage may be speeded up by kneading the powder into the 'putty' by hand. Precautions must be taken to avoid inhaling asbestos powder.

The use of a 2 per cent solution of silicone fluid or grease in ethyl acetate or any other suitable solvent as a separator between the mould and the specimen is advisable.

Mould boxes are not necessary when a mixture contains 50 per cent or more by weight of polyvinyl chloride powder. Casts from these moulds may be made in the paste itself, plaster or resins and fibreglass. Except when plaster is used, an appropriate separator is necessary. When polyvinyl chloride pastes are employed to make a cast either from a polyvinyl chloride, silicone rubber or a natural mould it is preferable that they should be coloured. The makers sell a range of colours which may be mixed into them but a brown material may be made by mixing a little burnt umber powder into some dibutylphthalate, straining this through a piece of nylon mesh into the paste and stirring it until the colour is evenly distributed; titanium dioxide may be used in the same way to colour the plastic white.

Preparation of Specimens for Moulding

Before commencing to make a mould from any specimen the surface must be sound. Any consolidation should be done with a substance which will not be affected by the moulding medium or the temperature at which it is used and any large cracks should be filled. Except when rubber latex or polyvinyl chloride pastes are the media, plasticine can be used. Modelling clay will not interfere with rubber, and the putty made from asbestos powder and silicone grease will stand up to the curing temperature of the polyvinyl chloride pastes. The water-soluble putty based on polyethylene glycol 4000 described in Chapter 5 is very convenient for filling cracks where silicone silastomers are employed. It is by far the easiest substance to remove from a specimen after moulding. The orbits, nares and other parts of skulls sometimes need to be filled to make the moulding simpler.

All labels should be removed from the specimen and stored safely to be replaced after the mould has been made. At all times a temporary label must remain with the specimen. If the fossil is inside a mould the label must be firmly attached to the outside.

Making Moulds

The simplest moulds are those made of fossils which are only partially exposed on one face of a piece of matrix. If rubber latex or a cold-cure silicone silastomer strengthened with surgical gauze or bandage is to be used the only preparation needed is to make sure that the specimen is sound and clean and to apply the appropriate separator. The moulding material is then brushed onto the specimen and the strengthening material added between each application.

When the required thickness has been built up and the mould medium has fully set, a backing of plaster of Paris is applied. This, in small specimens, may be spread on when it has set to a point which makes this possible or it may be poured on, in a more liquid state. Pouring requires that a dam should be built round the edge of the mould, for rubber latex, modelling clay should be used and this or plasticine will do for silicone rubber. Very large moulds require a large backing of plaster and to retain strength, and at the same time to reduce weight, the plaster may be reinforced with plasterer's scrim applied in strips while the plaster is still in a workable condition. The plaster should be left until it has fully set before it is removed from the mould. It is important to be sure that the plaster backing in this type of mould does not lock into an undercut. This can be avoided by packing out any which appear on the back of the mould with some of the moulding material filled with jute floc before the plaster is applied.

The majority of moulds, even those of one piece, are made by pouring the moulding compound onto the specimen. To hold the liquid in position until it sets it must be contained in a dam. This may be made of plasticine, clay or heat-resisting putty alone or these materials may be used as a seating for a wall made of thin flexible metal such as aluminium or brass. When the moulding material does not contain water, cardboard can be used. The strips should be formed so that the two ends overlap and can be held together with paper clips or adhesive paper. The gaps between the overlapping ends and the sides may be sealed with adhesive paper or the plastic material into which the bottom edge is set. Care must be taken to see that the bottom edge of the wall is set deeply enough into the plasticine or clay to stop the liquid leaking out. The soft material needs to be pressed tightly to the sides of the metal or cardboard with a modelling tool. Such dams may be built onto the matrix surrounding the fossil if the block is large enough. Flexible dental wax is very useful in some instances. When it is too small for this to be done, it should be embedded up to the level of the specimen in plasticine or its equivalent, and enough of this should be allowed beyond the edges to permit a rigid wall to be pushed into it and made secure. The surface on the inside of the wall should be made level and smooth. The whole of this process should be done on a flat piece of board, glass or asbestos sheet so that the set-up is easy to move. A dam may be made of gummed paper tape if the edges of the block containing the fossil are even; several layers are wrapped round so that they stick to the sides and to each other. This of course, cannot be used if the moulding material contains much water. The dam having been made, any separator required is applied.

The pouring of any liquid moulding material must be done in such a way that it enters all ornament and no air bubbles are trapped. It should always be poured from one point to the side of the specimen and not directly above it thus ensuring that the liquid will rise up the side and flow across the specimen. The advancing edge should be kept as near to a straight line as possible and should not be allowed to develop into a 'pincer' movement, because when the two jaws of the 'pincer' close, air will be trapped. To avoid the trapping of air in areas on a fossil which has ornament made up of a number of overhanging ridges, the liquid is poured so that it runs into the base of the ornament, fills the hollows and then flows out and over the edges of the overhang. The escape of air is helped if when the pouring is completed, the board on which the operation is carried out is tipped from side to side and gently tapped on the top of the bench.

The less rigid moulds may need supporting in a plaster 'Mother'. This is made by removing the dam from the sides of the set medium, extending the plasticine or its equivalent to a diameter slightly greater than the thickness of plaster required and making another dam into which the plaster can be poured. In practice a simple support of this type is just as efficient as an elaborately constructed mould box. The specimen is removed when the mould and the plaster jacket, where it is required, have fully set. The plasticine bed is freed from the board with a flat blade or a cheese wire and is carefully removed from the specimen which is then withdrawn from the mould.

Moulds of two or more Pieces

Making a mould of more than one piece is naturally more difficult. Using flexible materials the majority of invertebrate fossils can be moulded in two pieces, and this is true for most vertebrate bones. Multipiece moulds are not normally necessary except for skulls and the fused pelvic bones of some animals; this applies to very large bones as well as small ones. The partial skeleton of the sauropod *Cetiosaurus* has been cast from hot melt polyvinyl chloride moulds and none of these was made up of more than two pieces. The most important consideration when designing a two-piece mould is the position on the specimen through which the junction line of the two halves runs. It must be placed so that the fossil may be removed from the mould without damage and since it determines the position of the 'flash line' on the cast, it must run through areas of the specimen which show little or no ornament or are of the least anatomical importance. The 'flash' is that small excess of material which projects from any cast made from a mould of more than one piece. This has to be trimmed off and if it is too thick the cast is spoiled. In bivalves the 'flash line' can often be positioned along the natural junction of the two parts of the shell, although it may be necessary to avoid the area of the hinge.

Highly ornamental gastropods and ammonites present special difficulties. There is no set pattern which can be followed and each fossil has to be considered separately. Vertebrate fossils are somewhat easier for even when casting skulls, which normally require at least a four-piece mould, it is usually possible to arrange for the junction lines to run through areas of little morphological importance.

Having decided where the junction line of the mould is to run,

Moulds of two or more pieces 209

the first stage in its manufacture is to embed one side of the fossil up to this level in a material like plasticine (Fig. 23a). This must be

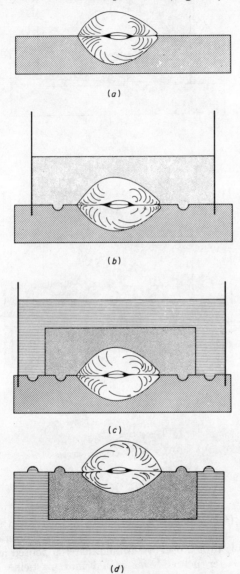

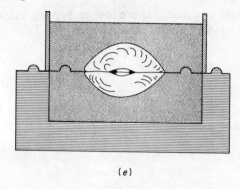

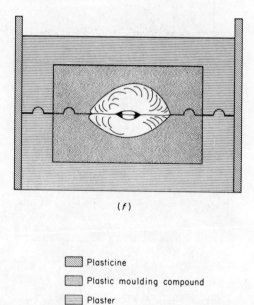

	Plasticine
	Plastic moulding compound
	Plaster

Fig. 23. Stages in making a two-piece mould. (*a*) Specimen half embedded in plasticine; (*b*) Moulding material poured into metal dam; (*c*) Metal dam repositioned and plaster poured on top of mould; (*d*) Plasticine removed and the whole turned over; (*e*) Mould material poured into plasticine dam; (*f*) Plaster poured over mould into plasticine dam; mould completed.

done very carefully and is the most time-consuming part of the operation of moulding and casting. The plasticine bed should extend beyond the fossil on all sides to accommodate the thickness of the walls of the mould and that of any plaster jacket which may be required. It must be worked tightly to the specimen all round and must terminate flatly and sharply against it. The whole bed, from the specimen outwards, must be smoothed with modelling tools and small steel spatulas; in particular no loose pieces or rough edges must be left at the junction of the plasticine and the fossil. When this has been done satisfactorily the round end of a hard-glass test tube, or ignition tube is first dipped in french chalk and then pressed into the plasticine in several places, at a distance of half the required wall thickness of the mould from the specimen (Fig. 23b). This produces a number of concave pits in the plasticine which will result in the formation of convex projections on the edge of the mould; for most moulds three or four depressions are enough.

A dam is now made outside the depressions at half the desired wall thickness of the mould from them (Fig. 23b). Generally a strip of metal or cardboard can be bent to shape and pushed into the plasticine. The junctions of the sides of the wall and the plasticine should be smoothed and a strip of plasticine moulded to the outside bottom edge to prevent any leaks.

A separator, when required, is now applied to the specimen and the plasticine within the dam; if it contains a solvent, time must be allowed for this to evaporate.

Taking the precautions already enumerated to avoid trapping air bubbles, the moulding material is now poured and allowed to set (Fig. 23b). When a plaster 'Mother' is required, the dam around the now solid mould is removed and the slit in the plasticine repaired and smoothed over. Suitably placed depressions in the bed are made with the hard-glass test tube and a dam made outside these. Plaster is now poured into the dam and over the half mould and left to set hard (Fig. 23c).

The plasticine bed is detached from its board with a flat knife or a cheese wire and the whole assembly turned over and with great care the plasticine is taken off to reveal the other side of the specimen in its half mould (Fig. 23d). All possible precautions should be taken to avoid moving the fossil in the mould.

The last traces of the plasticine are cleaned off the exposed side of the specimen and a separator, when indicated, is applied to it and the edges of the mould. A dam is now required into which the second half of the mould can be poured and if a mould box has been made for the first half the dam can be built up onto it to the edge of the plastic (Fig. 23e). When no box exists, the dam in the case of a small specimen can then be made by sticking adhesive paper round the outside of the plastic to project above the level of the fossil, or a metal strip may be fixed to the sides with elastic bands or masking tape. The second half of the mould is poured into this (Fig. 23e). When it has set, the dam is removed, and if a second half of a plaster box is needed, the edge of the existing half is given two coats of a solution of soft soap or shellac and more plaster poured into a dam formed round its edge (Fig. 23f).

On opening the two halves of the mould it will be seen that owing to the depressions made in the plasticine with the hard-glass tube, it is provided with a series of pits and projections on the edges which ensure the accurate positioning of one half on the other (Fig. 23d).

Polyvinyl chloride hot melt compounds will lose some of the oil they contain if they remain, for any length of time, in contact with untreated plaster of Paris, and for this reason moulds of this type should be removed from plaster 'Mothers' as soon as possible and the plaster given a heavy coating of shellac, which must be absolutely dry before the plastic is put back into the box. The same treatment is advisable when the mould is made of gelatine.

Sometimes thin-walled moulds tend to drop away from the edges of a plaster 'Mother' after the specimen has been removed. They may be pulled back into position by driving dressmaker's pins through the plastic into the plaster or by a piece of masking tape running from the edge of the mould to the edge of the 'Mother'. When this difficulty can be foreseen it may be avoided by modifying the dam into which the liquid moulding medium is poured, so that 'keys' are formed which will lock into the plaster 'Mother', but this is only possible when the dam is made of a malleable material, such as plasticine. At intervals along the base of the dam, trapezium-shaped holes are cut having their upper sides parallel to and twice the length of the lower; the other two sides being inclined at equal angles (Fig. 24a). Divergent walls of plasticine, which slope at the same angle are built outwards from the inclined sides, for a distance determined by the size of the 'key' required, and their ends are joined by a third wall, thus forming an enclosed space having the shape of a truncated isosceles triangle. The tops of these walls are now built upwards so that they form a chimney attached to the outer wall of the dam for its entire height (Fig. 24b). All junctions must, of course, be sealed to make the structure leak proof and the walls may be strengthened by building buttresses on their outer sides. When the liquid medium is poured into the dam, part will flow through the holes and up the chimneys; so after the medium has solidified and the dam is dismantled, at each of these points there will be a column of the flexible material separated from the main mass above the top edge of

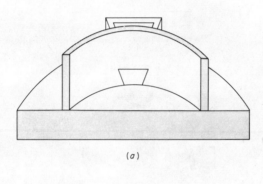

(a)

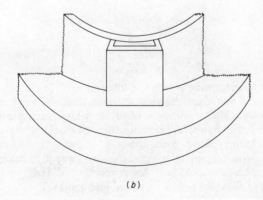

(b)

214 Casting

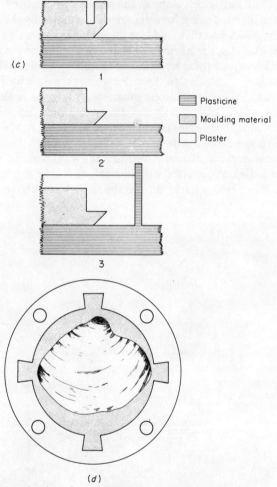

Fig. 24. Stages in forming keys in moulding material. (*a*) Section through inside of moulding dam showing aperture through which moulding material flows; (*b*) Outside view of (*a*) showing chimney into which moulding material flows; (*c*) 1, Section through moulding material after dam and chimney have been removed, showing excess moulding material; 2, Moulding material with excess cut away to form key; 3, Section showing how key locks into plaster 'mother'; (*d*) Half mould viewed from above to show shape of keys locked into 'mother'.

the trapezium-shaped holes (Fig. 24c1). The columns are cut through at this level, leaving truncated triangular 'keys' whose bases are farthest from the edges of the mould (Fig. 24c2).

The dam for the plaster 'Mother' is built so that it encloses the 'keys' and when the plaster is poured it will flow over them, and on its solidification they will be locked into the position and can only be withdrawn if they are deliberately distorted and prised out of their sockets (Figs. 24c3, 24d). 'Keys' for latex or silicone rubber moulds, made by brushing the media onto the fossil and then fortifying them with surgical gauze, may be made if the edge of the mould is extended beyond the edge of the specimen to form a horizontal 'brim', then cutting appropriate shapes from discarded moulds and welding these to the extension with liquid latex or silicone rubber.

Casting from Moulds

Plaster of Paris is the traditional substance used to make casts, and it is still the cheapest and most readily available. Opinions vary as to the best grade for this purpose, but it has been found that fine baked industrial or moulder's plaster is better than most. The general run of dental plasters tend to set too quickly and it is difficult to make a cast in them which is free of air bubbles. To mix plaster for casting it must be sprinkled evenly into water and this can only be done into a vessel with a wide mouth. Kitchen mixing bowls, made of plastic or porcelain, may be used provided that they have a spout to facilitate pouring; smaller amounts may be mixed in evaporating dishes, which are available in a wide range of sizes from suppliers of chemical apparatus.

The mixing of plaster and water has been described previously for such operations as the application of plaster bandages in the field. For making casts it must be done with more care; more than enough water to fill the mould is put into a basin and as for other uses, the plaster is sprinkled into it, but this time it should be done so that it is distributed evenly and finally settles to form a flat-topped layer which is approximately 2 mm below the surface of the water. An alternative and slightly easier method is to sprinkle the plaster into the water until it settles at about 1.25 cm below the surface and then pour off the excess water into another bowl.

A small quantity of the plaster will come over with the water, so it should never be poured into a sink. The stirring of the mixture must be done so that it forms a smooth lump-free mixture containing as few air bubbles as possible and for this reason a beating action should be avoided. No matter how carefully it is mixed the liquid will contain some air; this is dispersed, before the plaster is poured into the mould, by banging the bottom of the bowl up and down on the bench top and slapping the sides with the hands, a technique which will be recognised by any cook as very like the method used to force air out of some cake mixtures. The air will appear as bubbles on the surface and they can be skimmed off with a piece of cardboard.

Casting from one-piece Moulds

Casting from a one-piece mould is a very simple process. The mould is placed on a small piece of board and a separator is applied. After this has been done the surface should be examined to make sure that the brush has not shed any hairs and that there are no other foreign bodies present. The amount of separator used should be as little as possible since any excess may fill up detail in the mould; for this reason a silicone oil or grease is to be preferred as they are efficient in very small quantities. They may be applied as solutions in solvents but can be used alone in cases in which it is known or suspected that a solvent will damage the mould. When used in this way a brush is dipped into the silicone and the hairs are then wiped almost dry before they are applied to the mould. Several casts can be taken from a silicone-treated mould before the coating has to be renewed, but some experience is required to know just how many. All this should be done before the plaster is mixed.

When pouring plaster into a mould the same precautions against the entrapment of air as were described for pouring liquid moulding compounds should be taken and the same considerations concerning the filling of ornament also apply. Sometimes it helps to tilt the mould slightly towards the basin in the early stages.

The filled mould is rocked from side to side and tapped on the bench to force out trapped air. A plaster cast should not be taken from a mould in a hurry. The heat generated by the setting must be allowed to disperse and the cast should be left in the mould for a quarter of an hour after it has become cold, for if it is removed before this, it is likely to be damaged. The cast will take some time to dry; how long of course, depends on its size but nothing should be done to it until it is dry.

Strengthening Plaster Casts

Untreated plaster casts are very susceptible to damage. The plaster is brittle and easily broken and the surface can be abraded. They can be strengthened in several ways. Heating in various mixtures of waxes used to be common practice. It added some strength but if a cast treated in this way is ever broken, it is very difficult to repair, as the wax inhibits all the best adhesives, and about the only way to mend it is to use some of the original wax in a molten state. Impregnation with solutions of plastics is an improvement, but the best results are obtained if a cold-setting polyester or epoxy resin is used. These must be of low viscosity and the accelerator, catalyst or hardener proportions adjusted to give a pot life of about an hour, otherwise an excess of resin may harden on the outside of the cast before it can be wiped off. Exact proportions cannot be given as the setting time depends on ambient temperature, but a 1 per cent accelerator and 2 per cent liquid catalyst mixture should be satisfactory in most cases where a polyester is used. Manufacturer's technical leaflets are the best guides for the proportion and type of hardener to use with epoxy resin. Certain epoxy resins are used to consolidate sand moulds in metal foundries.

Colouring Plaster Casts

Plaster may be painted with dry colours, using a thin solution of Butvar B98 or another plastic as a binding medium, with acrylic colours, or the stains used in microscopy. Oil colours are not recommended as they tend to obscure detail. Almost anybody can, with practice, produce a good result with dry colours as they can be mixed to provide a wide range, can be applied one on top of the other and can be washed off without damaging the specimen.

Flexible Casts

Flexible casts in polyvinyl chloride may be taken from moulds made from fully cured polyvinyl chloride pastes with a solid content of 50 per cent or more. In the fully cured condition, such a mould when cold, can be filled with some liquid paste and after the air has been forced out by tapping it on the bench top and allowing it to stand for one hour, it can be placed in an oven and raised to 150°C once more; the liquid filling will solidify, and when the mould is cold it can be removed as a flexible cast. The separation of the mould and cast is easier if it is done under water. Although theoretically no separator is required, it is safer if a little silicone is used.

If a fossil, such as a trilobite, is resting on a bed of matrix a cast differentiating in colour between the animal and the rock may be made. A quantity of paste, coloured by mixing burnt umber into it is carefully poured into the mould up to the level of the junction between matrix and fossil and is raised to 90°C and then allowed to cool. Some more paste coloured white with titanium dioxide or grey with slate powder is poured on top to fill the mould, and the whole is then raised to 150°C and cooled. The result is a very durable cast strong enough to remain undamaged even if thrown against a wall; it shows great detail and is extremely useful for teaching purposes. The property of solidification at 90°C can be used to ease the making of casts from complicated natural moulds, as one part may be filled and solidified and another which joins it at an awkward angle may be filled with the liquid and the whole raised to 150°C when the two pieces will weld together.

Solid Resin Casts

Solid polyester or epoxy resin casts can be taken from polyvinyl chloride or silicone rubber moulds and, with due care, from gelatine, but there is seldom any point in doing this from a one-piece mould as a fibreglass cast is so much lighter and stronger and so very easily made.

Fibreglass Resin Casts

For faithful reproduction of fine detail, strength and lightness, casts in fibreglass are without equal. The technique is not difficult and where single-piece moulds are concerned, even the first attempt should be quite satisfactory. Moulds in polyvinyl chloride hot-melt compounds should be dusted with talc. Silicone silastomers require no separators. Polyester resins are more generally used than the epoxy resins although the latter have no shrinkage. There is no real difference in technique except that the epoxy resins normally require the addition of a hardener only to activate them. Some polyester resins can be bought with the accelerator ready mixed in and these require the addition of a catalyst only to cause them to set. If a polyester resin is to be used in large enough quantities and in a reasonably short time, one may always save time by mixing the accelerator into it in bulk. Some manufacturers produce polyester gel coat resins which are pre-accelerated and others do not, but provide a pre-gel paste which has to be mixed in a stated proportion with the ordinary resins. All manufacturers of both polyester and epoxy resins issue technical leaflets and are usually pleased to supply these to serious users of their products. Technical leaflets and manufacturer's handbooks are an essential part of the equipment of an institution producing fibreglass casts, for while it is possible in a book such as this to outline general principles it is not possible to discuss every resin which is made. Those which are mentioned are chosen only because the writer has had experience with them; but they may not be easily obtainable all over the world and so other brands may have to be used. The resin chosen should be of low viscosity, have high resistance to chemical attack and be recommended by the maker for use with fibreglass.

Glass fibre casts are becoming increasingly popular and this makes it wise to include a fire retarding agent in the mixture, for although the glass does not burn and even acts as a fire barrier, the resin which impregnates it will burn very easily unless a retardant is used. Most manufacturers supply such chemicals and will give advice as to the best ones to use.

The catalyst for polyester resins is supplied both as a paste and as a liquid. The paste is safer but is more difficult to use, since it has to be weighed for each batch of resin and then has to be very carefully stirred in to ensure even dispersal. The liquid is an organic peroxide which is corrosive to the skin and is a fire hazard, since it can cause spontaneous combustion if it is allowed to come in contact with sawdust, paper or cloth. It must never be stored in bulk inside a building but must be confined to a properly constructed store for inflammables and the fire staff told that it is there.

Working quantities should be kept as low as possible and the container always placed in a porcelain tray which is large enough to retain more than the contents of the bottle should it break. It is best dispensed from a burette or a broad-based bottle fitted with an automatic pipette. Neither should be of large capacity and should always remain in the porcelain tray. If a burette is used, the stand must not be made of wood and must be firmly fixed so that there is no possibility of it being knocked over. On no account should the accelerator and catalyst, either in paste or liquid form, be mixed together as the reaction is extremely violent and dangerous. For this reason the two should be at all times be kept well apart so that there is not the remotest chance of accidental contact occurring.

The choice of catalyst will be that of the user, but it is strongly recommended that when the output demand is not high it should be a paste. Some of the above has already been said in this book, but it is worth repeating. A hazard is less likely to develop into an active danger if all the facts are known and commonsense is applied.

The glass fibre normally used for casts of this nature is of two types: a surfacing mat which is very thin and a heavier chopped strand mat. Loose chopped glass fibre strands are also available and those from a quarter of an inch (6 mm) to one inch (2.5 cm) in length may sometimes be used, with advantage, in place of chopped strand mat. Details of types available can best be obtained from the manufacturers. Surfacing mat is graded according to thickness and where there is a choice, the thinnest available should be used. Chopped strand mat is sold in various weights per square foot (approx: 0.093 square metre). For use in casting and other palaeontological techniques, one having a weight of 1 oz per square foot (approx: 300 gm per m^2) is suitable.

The first step in making a fibre glass resin cast is the application of a gel coat to the mould. It is advisable that this should be coloured because it is then easier to see that it has been applied evenly and no voids are present. The resin may be coloured by the addition of approximately 3 per cent of yellow ochre powder or titanium dioxide, dispersed evenly throughout the mixture, dispersal is assisted if the colouring matter is first mixed thoroughly into a small quantity of the resin to make a lump-free paste and then stirred into the remainder. At normal room temperature it is not wise to attempt to use more than 100 g of resin at any one time as it sets quickly. When using a preaccelerated polyester such as Trylon Gel Coat Resin G.C.150 PA, the required amount is weighed out in a paper cup and after it has been coloured, as described above, 2 per cent of the liquid catalyst is added and stirred in very thoroughly. It is then brushed over the inner surface of the mould. Because it is thixotropic the gel coat will flow easily under the friction of the brush but as soon as the brushing ceases it will gel and remain, even on vertical surfaces. The coating of large moulds will require more than 100 g of gel coat and in these cases the first 100 g batch is brushed onto as large an area as it will cover efficiently and then another batch is mixed and applied as before, overlapping the edges slightly to ensure a complete junction. This process is repeated until the mould is covered. If necessary a second coat may be painted on after the first has set.

While the gel coat is hardening some surface mat is cut into small pieces; their size depending on the contours of the mould. The more complex the surface of the mould, the smaller the pieces need to be, for if they are too large there is a danger that the action of stippling them into one part of the mould will pull them out of another. It is advisable to cut up more surfacing mat than you expect to use because the activated resin used from now on has a short pot life and if the supply of mat runs out before you have covered the mould surface it results in a waste of materials and time.

The resin used to fuse the surface mat to the gel coat and later the chopped strand mat to the surface mat must, be capable of saturating the glass fibre and therefore has no thixotropic agent mixed with it and is of low viscosity. Trylon W.R.180 fulfils these conditions and a fire retardant may be mixed with it. For use in the following stages of making the glass fibre cast, Trylon W.R.180 is mixed with 2 per cent of accelerator, 15 per cent of the fire retardant Prefil 'F' and 2 per cent of liquid catalyst, in that order, and the whole stirred thoroughly.

A small area of the hardened gel coat is painted with a little of this mixture, a piece of surface mat is placed on top and is stippled down with a stiff brush loaded with the same resin mixture. It is important that all the air is forced out from between the mat and the gel coat. This process is repeated with every piece of mat slightly overlapping the one before it, until the whole mould surface is covered and if this has been done properly the pieces of mat should coalesce. A second layer of surface mat may be applied but this is not always necessary. After the resin has set, the final layer or layers of chopped strand mat are applied. Enough to cover the mould surface is cut into convenient sized pieces before any resin is mixed.

Chopped strand mat is very much thicker than the surface mat and although it can be applied directly by stippling with a brush loaded with resin by a practised operator, a beginner may find it easier if it is first saturated with resin. This can be done on a flat surface covered by cellophane. The piece of mat is placed on this and some resin painted onto it; it is then turned over and more resin applied on the other side. Then the resin-soaked mat is lifted on a spatula, placed in the mould and stippled into place. The process is continued until the mould surface is covered. The individual pieces of mat should overlap each other slightly and should coalesce. One layer of this mat is often enough but as many as may be required can be built up.

The resin is left to harden and the cast removed from the mould; the rough edges which project beyond the mould are trimmed off. The trimming is done with scissors or a hack-saw blade or by grinding with 'Abracaps' or some other type of rotating abrasive burr. Glass fibre resin casts can be painted with power colours or with acrylic paints. The resin has to be weighed and the best balance for this is one of the beam types with sliding counter weights. A balance with a rail fitted with a tare weight, to cancel out the weight of the cup, is useful.

Between the time that one cup of resin is finished and the next is mixed, the brush, if left alone, will be rendered useless by the resin solidifying in the bristles, so it must have the excess resin squeezed out of it and then be rinsed and left standing in acetone until it is needed again. When all the resin has been applied the brushes are washed in acetone and then in concentrated liquid detergent which in its turn is rinsed out in water. An extravagant, but efficient alternative method is to remove the brush from acetone and clean the bristles with a resin-removing cream intended originally to remove resins from the operator's hands.

At all times when handling resins the worker should wear rubber gloves and should be provided with a barrier cream for use before and after the job, for if this is not done there is a danger of dermatitis.

Glass fibre casting should never be done in a badly ventilated room; equally, it should not be done in a draughty environment as this can lead to an uneven curing of the resins. In the liquid stage, every ingredient of a polyester resin is inflammable so no naked flames can be allowed in the working area.

Casting from Moulds of more than One Piece

The making of casts from moulds of two or more pieces requires much skill and practice. The most important thing is to keep the junction between the separate pieces of the cast, the 'flash line', as thin as possible; ideally it should not appear on the cast at all, but in practice it cannot be avoided.

It is obvious that to approach as nearly as one can to the ideal state, the parts of the mould must close accurately and there must be no gap between them. For this to be so the faces of the mould and any surrounding 'Mother' which come into contact with each other must be absolutely clean and free from foreign bodies. Particular attention should be paid to the concave pits which are there to ensure a true register, for foreign matter can easily get into them and even if only one of them is fouled when the mould is closed the cast will be useless. It is also of prime importance that once the mould has been closed it is held tightly shut. This may be done in a variety of ways, ranging from placing weights on top of the 'Mother', the use of elastic bands, and tying with string or rope. Powerful elastic bands, which are invaluable, may be made by cutting sections out of a discarded car tyre inner tube. On large moulds, mainly for glass fibre casts, screw clamps can be used if the plaster 'Mother' is designed to take them. In archaeological laboratories, moulds made of silicone silastomers and bandage are often supported in 'Mothers' made of glass fibre and resin. These are equipped with a rim which is reinforced in places through which metal bolts are passed and closure is achieved by screwing wing nuts onto the bolts. This method is very efficient, but the 'Mothers' must be made very accurately.

Plaster Casts

There are two ways of making a cast in plaster of Paris from a two-piece mould. Both are limited to small or medium-sized objects and very few people succeed on the first attempt at either. The first method is to place the two halves of the mould, when they have been properly cleaned and when necessary lubricated, side by side and pour into each half slightly more mixture than is needed to fill it; the excess should stand proud of the mould edges as the reverse of a meniscus. The plaster is left until it has thickened to a point where one half of the mould can be inverted and pressed down onto the other. This requires rather nice judgement, because if it is done too soon the plaster will drop out and if too late the two halves will either not press together and join up or do so and produce a flash line which is far too thick. A rough method of gauging the exact point is to keep an excess of the plaster in the mixing bowl and from time to time take up approximately as much as is in the half mould on the end of a spatula or in a shallow spoon, and when this can be inverted without immediately falling, the plaster in the mould is ready.

The mould having been successfully closed must be held together. This is most easily done by putting a weight on top of it, but this must be placed so that the pressure is evenly distributed. If there is no 'Mother', as in the case of silicone silastomer or polyvinyl chloride paste moulds, the weight must be placed on a piece of wood which more than covers the top of the mould. It is obvious that the weight must not be heavy enough to cause distortion.

The second method of making a cast from a two-piece mould is to fill one half of the mould with plaster mixture, close it and 'spin' it; which means that the mould is rotated in all directions so that the liquid plaster flows evenly over all surfaces. The mould is held closed by string or elastic bands and the rotation must continue until the excess plaster in the mixing vessel has set hard. The mould should remain closed until the plaster in the bowl is cold. This method produces a hollow cast and if the mould has closed properly, the flash line should be very faint.

A variation on the first method is sometimes used to produce plaster casts of large specimens. This requires that the mould be made in three pieces, two large and one small. The two large pieces are fixed together and slightly overfilled with plaster and the same is done with the small piece. When the plaster is in the right condition to make the inversion of the small piece possible this is done and it is put in place and the two partially set areas of plaster are pushed together.

The pouring of plaster into a closed mould is seldom possible when casting fossils. It can only be done with specimens in which there are conveniently placed broken areas or pieces of matrix. In such cases, the mould and any 'Mother' is made so that holes run through them into such areas. One hole must be large enough to allow a plaster mixture to run in and the other is there to allow the displaced air to escape; if there are several areas which are barren and the specimen is large, more than one vent hole can be made. Should these be below the upper level of the liquid plaster they may be plugged with plasticine or clay as it reaches them and starts to leak out. The resulting columns of plaster on the cast are cut away and the scars may either be left or obliterated by carving the zones to match their surroundings.

Flexible Casts

Flexible polyvinyl chloride casts can be made from polyvinyl chloride paste moulds using the same material. Each half of the mould is treated with a silicone separator and very slightly less of the plastic than is required to fill it is poured into each. The mould is tapped up and down on the bench and left to stand for an hour to allow air bubbles to rise to the surface and to disperse. Both halves are then placed in a controlled electric oven set at 90°C and left until the plastic sets; they are then removed and allowed to cool; enough plastic is painted with a brush into each half to slightly overfill it and finally the mould is closed and held together with adhesive paper tape, elastic bands or small weights. It is returned, together with a tube of paste, to the oven which is now set at 150°C. When the paste has set to a strong elastic solid the mould is removed from the oven and cooled; which is better done in air rather than by plunging it into water.

On opening the mould the two halves of the cast should have welded together and there should be a very thin 'frill' of the plastic as a 'flash line', which can often be pulled off without leaving a scar; if it will not break it can be cut away with small scissors.

A silicone silastomer cast may be made from a gelatine mould in much the same way but no heat is required. The two parts are once more slightly underfilled and when the silastomer has set some newly mixed material is painted onto each half and the mould closed.

Solid Resin Casts

The same technique of slight underfilling, allowing the material to set and then welding the two halves with some fresh liquid is used to make solid resin casts from any suitable mould. The 'flash' on casts made this way is very thin and easy to remove; for this reason this method is very good for the reproduction of small ammonites, brachiopods, mollusca and small teeth of vertebrates.

Glass Fibre Resin Casts

Glass fibre resin casts can be made from moulds or two or more pieces. Each piece is filled in the same sequence of operations as described for use with a one-piece mould. Time can be saved if any overlap of the glass fibre beyond the edges of the mould is kept as small as possible, because this must either be cut or ground off flush with the edges or the mould will not close accurately. The two halves are joined by first applying to each a series of overlapping strips of the finest surfacing mat, so that part of them is flat to the mould edge and part is applied to the adjacent inside surface of the glass fibre. These strips must lie absolutely flat on the plastic which extends beyond the edges of impression of the specimen and the amount of resin should be as little as is practical

(Fig. 25). When the resin which impregnated this border of glass fibre has set, some of a newly-mixed batch is painted onto each side and the mould closed. To ensure that enough resin remains on the exact junction line of the two halves, the resin used to weld them together may be a gel coat and some should be painted inside the body of the two halves in this area. It is weighted or held tightly shut by some means until at least twice the length of time it takes the excess resin in the mixing cup to set has elapsed. When the mould is opened and the cast removed the parts should have joined and the 'flash' of fibreglass can be cut away.

An alternative method is to pour some liquid resin into one half after the edges of the glass fibre resin has been trimmed flush with those of the mould, close the mould and rotate it so that the liquid will flow along the junction line. The rotation must continue until a control sample of the mixture has set. These two methods may be combined. It takes practice to get good results this way, but since the surfacing mat used is very thin and is further flattened by stippling in the resin, it is possible to produce casts without an offensive flash line.

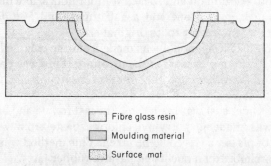

Fig. 25 Section through half mould showing position of the strips of surface mat used to join the two halves of a fibre glass cast. For clarity thickness of surface mat greatly exaggerated.

Enlarged Casts from Silicone Silastomer

The following method has been used to make enlarged casts of fossil fruits for exhibition. The originals were too small for the detail to be seen without the use of lenses. It was developed by

Mrs Jane Croucher and the author and although first used for fruits it could most usefully be applied to any small fossil and the results are better than models.

The fossil was first moulded in gelatine.

The specimen was half embedded in asbestos and silicone grease putty, but plasticine could have been used, and surrounded by a dam made of aluminium strip. A 2 per cent solution of silicone fluid in carbon tetrachloride was applied to the fossil as a separator. When the solvent had evaporated the gelatine was poured and allowed to set. The solid mould was turned over, the embedding material removed from the fossil, the separator was applied and a dam of adhesive paper tape made round the mould. The second half of the mould was poured at about 40°C to avoid remelting the first half.

Each half of the mould was almost filled with silicone silastomer mixed with catalyst. When this had almost set a thin layer of newly mixed silastomer was applied to each half, the mould closed, and time allowed for the silastomer to set. The cast was removed and the flash, when present, was trimmed off and then it was placed into paraffin oil and soaked for four hours, in which period its volume increased one and a half times, and 5–6 days were required for it to revert to its original size.

A second gelatine mould was made from the enlarged cast and another silastomer cast was taken from it. This was enlarged in paraffin oil (kerosene) and the process repeated until a cast large enough to show the detail was obtained.

A silastomer cast was made from the last mould and from this a mould was made in polyvinyl chloride paste from which any number of plaster casts could be taken. This method was found to be more reliable than one using natural rubber latex.

References

Clarke, C. D. (1938) *Moulding and casting*, Butler, Maryland: Standard Arts Press.

Fischer, A. (1939) 'Rubber cast and moulds of fossils', *Journ. Paleont.*, **13**, 621.

Fuehrer, O. F. (1939) Liquid Rubber as an enlarging medium. *Mus. News*, **16**, 8.

Hoskins, J. H. and Cross, A. T. (1941) 'Techniques in the study of fossil plants', *Trans. Ill. Acad. Sci.*, **34,** 107–8.

Percy, H. M. (1962) *New materials in sculpture*, London: Alec Tiranti.

Quinn, J. H. (1940) 'Rubber moulds and plaster casts in the paleontological laboratory', *Tech. Ser. Field Mus. Nat. Hist.*, Chicago.

Quinn, J. H. (1952) 'Concerning rubber moulds and plaster casts', *News Bull. Soc. Vert. Palaeont.*, **36,** 28.

Rixon, A. E. and Meade, M. J. (1956) 'Casting techniques', *Mus. Journ.*, **56,** 9–13.

— (1960) 'Glass fibre resin casts of fossils', *Palaeontology*, **3,** 124–6.

Stahl, E. (1951) 'A new casting method for palaeontological purposes', *Bull. Geol. Inst. Univ. Uppsala.*, **34,** 201–3.

Vinamould hot-melt compounds handbook, Carshalton: Vinatex Ltd.

10
The Equipment and Management of an Ideal Laboratory

In Chapter 1 the basic essentials concerning ventilation, fire precautions and the avoidance of toxic hazards, to which any room used as a laboratory must conform, were discussed, but little else was said because it is only seldom that major alterations can be made to an existing laboratory. Occasionally a university or a museum may extend its premises, and then the laboratory staff may have the opportunity of acquiring at least some of the amenities they would like.

Location of the Laboratory

There may not be a free choice as to where in the building the laboratory is located, but there are certain factors which must be borne in mind when deciding where it is best placed. The most important are: the ease with which specimens coming in from the field can be conveyed to the laboratory and that the floor loading is well within safety limits. For these reasons a large laboratory expected to deal with heavy specimens and equipped with heavy machinery and furnishings should be at ground level, whenever this is possible.

Lighting

The preparator's work makes good lighting essential so the laboratory should have as many large windows as is architecturally practical, as well as provision for artificial lighting well above the level required in normal places of work. This is best provided by colour matching fluorescent tubes and considerable care should be taken in their arrangement.

Benches

Each worker must have his own bench and as many as possible should be placed below windows and the others against walls with overhead lighting. The bench tops need to be strong and made of a hard wood such as teak; below them at one end there should be a cupboard large enough to house a binocular microscope with its stand and the associated lamp and transformer; at the other end a nest of drawers in which the preparator may keep his tools and any small specimens upon which he is working. Consolidating solutions, adhesives and solvents are in constant use and a nest of shelves is required for their storage, these may be fixed either at the ends of the benches or to the walls.

At least four electric outlet sockets, rated at 13 A are needed at each working point. Units are now available fitted with electronically operated overload cut outs, which may be plugged into one socket and provide four outlets. These are very useful but it is advisable to consult the authorities responsible for the electric wiring of the laboratory before putting them into operation.

A gas supply is desirable but care must be taken in the location of the outlets. They must be positioned so that there is no danger of them being broken off. A means of turning off the main gas supply to each bench is a useful safety measure. A small sink placed at one end of the bench with a water supply delivered by a swan-necked tap fitting is a great convenience, in order to keep out chips of rock the sink should have a removable cover. Compressed air and vacuum pipes may be considered as a luxury, but when funds will run to it they should be installed.

As well as working benches others are required to accommodate thermostatically controlled electric ovens and ultra-sonic tanks; storage cupboards may be built below them. The benches so far described are only suitable for the preparation of small to medium sized specimens; vertebrate and invertebrate specimens enclosed in a considerable weight of rock have to be dealt with and it is necessary to have some benches capable of carrying a load of up to half a ton. It is impossible to forecast how large a bench may be required for this purpose. A good plan is to have a series of tables constructed all of the same dimensions, so that they may be placed together to form a working surface which is adjustable within wide limits of length and breadth. Tables of heavy construction 2 m (6 ft) by 1 m (3 ft) and about waist-high to an average man are invaluable in this context.

Large Sinks

As well as the small bench sinks mentioned above, the laboratory, depending on its size, should have one or more large deep sinks with a water supply, the outlets leading into sedimentation tanks. The drainage pipes must be made of a chemical-resistant material such as polythene. Hot water heaters and apparatus for the production of de-ionised water may be placed above them.

Fume Cupboards

The need for these was stressed in Chapter 1. As many as can be fitted into the space available should be installed and each should have internal lighting. Advice concerning the placing and construction of electric points should be obtained from an expert as there is the risk of corrosion effecting the wiring. Gas and electric outlets are best controlled from the outside. When ordering, it is important to specify that the ducting must be resistant to corrosive fumes and that inflammable vapour will be present, which makes it necessary that the fan motor be placed so that sparks from the commutator cannot ignite them.

Extraction Hoods

Because of the constant use of volatile solvents and the creation of finely divided dust by some methods of preparation, an extraction hood is desirable at each working point. The type attached to extendable and retractable ducting which allows them to be pushed up out of the way when they are not needed are the best. Such hoods are particularly necessary when large amounts of polyvinyl chloride hot-melt compounds are used and large fibreglass casts made.

Storage Cupboards

Besides those which may be built below benches, other cupboards are needed for the storage of raw materials and tools other than those forming the personal kit of each preparator. For most purposes these can be any form of wooden cupboard of convenient size, but metal units are required for inflammable solvents. The stock held in them should be as low as is practical and the doors must be labelled 'Contents inflammable' in large red letters; they should be placed well away from any fire exit. The bulk stock of inflammable materials should be housed in a specially constructed store outside the main building.

Refrigerators

Polyester resins, the accelerators and catalysts used with them, have a longer shelf life if they are kept cold. They may be stored in a refrigerator, but it must be established that it has no electrical components in its interior which may cause a spark. As with inflammable solvents, the amounts kept in the laboratory should be minimal and the refrigerator placed away from fire exits.

Metal Working Area

The chapter on mounting specimens for exhibition should have made it clear that the laboratory needs space in which metal work may be done and a supply of the appropriate tools. It is convenient if this operation is confined to an area partitioned off from the main laboratory. A fitter's bench complete with a vice is essential, a portable heavy duty vice with a folding stand is also required. Much time may be saved if a small bench top bandsaw complete with blades, capable of cutting wood, plastics and metal is available. For the forging of chisels and other tools, a gas forge, an anvil, a set of blacksmith's tongs and hammers are necessary. There should also be a rack for holding a supply of all the most commonly used metal bars, rods and tubes. Set of taps and dies for BA and Whitworth threads together with a good supply of twist drills of appropriate sizes, nuts and bolts both cheese headed and countersunk are needed. A hand operated 'pop' riveting machine, with interchangeable chucks, and a supply of the rivets used with it is a tool which can save many man hours. An electric drill, preferably of variable speed, fitted to a stand with a drilling table is absolutely essential.

Provisions for large-scale Acid Preparations

Ideally, institutions which carry out acid preparations on a large scale should have a special room built for this purpose. It should contain two large brick-built vats lined with an acid proof material with drainage holes at the bottom connected by polythene piping to an external sedimentation tank. In cases where hydrochloric acid is likely to be used in large quantities the advisability, or even the necessity, of installing a neutralising tank should be discussed with the local drainage authority. It is so obvious, that it scarcely needs to be said, that hydrofluoric acid should never be used on such a scale, for it would constitute an insurmountable health hazard to the staff, no matter how efficient the fume extraction system and it would be criminal to introduce the un-neutralised waste into the drainage system. This acid cannot be neutralised safely with anything but a large excess of calcium hydroxide, producing calcium fluoride which is insoluble in water and the disposal of this in any quantity would be a problem. The vats must, of course, be covered with an extraction hood sealed to their edges on three sides, with its fourth side hinged and fitted with transparent windows. The extraction system must be very powerful and the fumes passed through a coke filled washing chamber before being released into the air. When the vats are not in use for the preparation of large specimens, strong trestle tables treated with epoxy varnish may be placed in them and used to support smaller vessels, thus converting the extraction hood into a very large fume cupboard.

There should also be a fume cupboard in the acid laboratory in which hydrochloric and hydrofluoric acids may be used in small quantities. The room should be furnished with benches below the windows and against the free walls. Two large deep sinks are needed with acid proof plumbing and a water supply. In spite of the extraction system some acid fumes will accumulate in the air, so all electrical and gas fittings need to be designed and positioned with this in mind. For the same reason the windows should be fitted with extraction fans and the staff instructed that if they think the level is dangerous the rule is 'When in doubt, get out'.

Management

The chief preparator in a large laboratory has, often to his disgust, other duties besides the actual preparation of specimens. He is responsible for recording the reception of specimens into the laboratory and their safe return to their proper place after preparation, all technical correspondence, keeping an eye on expenditure to see that it does not exceed the laboratory grant, allocating work to other members of the staff and trying to obtain the highest output of work consistent with efficiency. He, therefore, needs a small office cum laboratory of his own. This means that besides adequate bench space he must have an office desk and filing cabinets. His most difficult task is ensuring that the laboratory does not become overloaded with specimens awaiting attention, for this not only endangers the specimens but has a demoralising effect on the staff.

One method which has been found to be fairly successful is to have a 'request for work' system operating. The theory is that when a scientist requires a specimen to be prepared or cast he fills in a card as shown in Fig. 26 and sends this to the chief preparator. When, and only when, a man is available to do the job the specimen is brought to the laboratory. The preparator to whom the work is assigned then receives the card and keeps it with the specimen until the job is completed. He then records on the back of the card what treatment the specimen has received, the date on which he started the job, to whom he returned the specimen and the date. The card is then filed under the zoological group to which the fossil belongs and is kept as a permanent record. Besides controlling the flow of work in and out of the laboratory this system has the advantage that should the fossil ever be returned to the laboratory, the card can be extracted from the files and the preparator can see at once what previous treatment it has received. Examination of the specimen will reveal whether this was effective or not and provides useful information on the long term behaviour of consolidants and adhesives; should it be necessary to remove the material the preparator knows exactly with what he is dealing.

SPECIMEN(S) No	GROUP	WORK REQUESTED BY	DATE
NAME OF SPECIMEN	TREATMENT REQUIRED (TICK WHERE APPROPRIATE)		
	DEVELOPMENT		CASTING
	REPAIR		MOUNTING
	CONSOLIDATION		EMBEDDING
LOCALITY	ARREST OF PYRITE DECAY		SECTIONING
	ANY OTHER TREATMENT		
	CONDITION OF SPECIMEN ON RECEIPT		
FORMATION			
	RECEIVED IN LABORATORY BY		DATE
NUMBER OF SPECIMENS	RETURNED TO	BY	DATE

Fig. 26. Request for work and laboratory record card.

11
Some Chemicals and Natural Substances used in Palaeontological Techniques

The descriptions given in this chapter are only the basic elements of the subject and are concerned solely with those properties which are of use to the palaeontologist. Whole books have been written on the solvents used in industry and these and standard works on chemistry should be referred to for fuller information. The manufacturers of industrial solvents publish information concerning their wares and the larger suppliers of chemicals do the same, some of them also give the name of the substances in several languages.

The information here is given in good faith but it is only fair to point out that the author is neither a qualified chemist nor a doctor of medicine. Boiling points are given as approximations since the exact value differs with the purity of the liquid.

Apart from certain palaeobotanical techniques it is not necessary to use pure chemicals for our purposes and commercial or technical grades, which are much cheaper, should be bought.

Whenever finances and storage space will allow, it is much more economical, and very often a great deal easier, to buy the more commonly used materials in bulk.

In these days of giant commercial enterprises it is becoming increasingly difficult to obtain supplies in small quantities, for large manufacturers nearly all have a minimum order, either as to weight or price, which they are willing to accept.

Details cannot be given here as they change with the sales policy of the firms and are sometimes effected by mergers or take-overs. Most of them are very generous about supplying technical details of their products and some will supply small samples for test purposes. There are however firms which cater for the small user and some of them are listed in Appendix II of this book.

The arrangement in this chapter is alphabetical for easy reference. The substances are merely divided into organic liquids and solids, inorganic liquids and solids and cold-setting compounds.

Organic Liquids

Acetone

Dimethyl ketone, colourless with a pleasant and characteristic odour said by some to resemble that of apples. Miscible with water. Solvent for cellulose nitrate, cellulose acetate, polymethylmethacrylate, polybutylmethacrylate and many other plastics. Both the liquid and its vapour are inflammable. Boiling point low, about 57°C. Toxic hazard due to vapour low, as great discomfort is experienced before a dangerous level is reached.

Acetic Acid, Glacial

Colourless with a very pungent odour reminiscent of vinegar. Used in 10–15 per cent (v/v) solution in water to break down limestones and other rocks cemented with carbonates, which contain fossils preserved as phosphates or silica. Boiling point above that of water but of no interest to the palaeontologist. Has a high freezing point, about 16°C, and solidification in cold weather should be guarded against. The glacial acid is corrosive to the skin. Should be used in a fume cupboard or in a very well-ventilated space as even at low concentration the fumes are unpleasant.

Amyl Acetate

Colourless with a very powerful odour of peardrops. Not miscible with water. Used as a high boiling point solvent to slow the drying of varnishes and adhesives and to stop 'blushing' due to the deposition of water in plastic coating as it dries. Boiling point above that of water, about 140°C. The liquid is inflammable. Will dissolve cellulose nitrate but not cellulose acetate. Should not be used in badly ventilated rooms as the vapour is irritant and causes headaches in some people. Can usually be replaced by *n*-butyl acetate which is not so unpleasant.

Benzene

Except where toluene cannot be substituted, which is very rarely, if ever, do not use this solvent. The physiological hazard is very great.

Bromoform

Colourless heavy liquid with an odour rather like that of chloroform. When pure it is colourless and has a specific gravity of about 2.9. It will yellow on exposure to light and air. That normally on sale has a small percentage of alcohol added, which reduces the specific gravity to about 2.65. Immiscible with water. Used in the flotation method for separating small specimens from a finely divided matrix. Non-inflammable. Both liquid and vapour are toxic. Boiling point high, about 149°C.

n-Butyl Acetate

Colourless, immiscible with water and having much the same odour as amyl acetate but less penetrating. Other properties and uses much the same and in almost all cases may be used to replace amyl acetate. It is considered to be a safe solvent as far as long term toxic effects are concerned. Inflammable. Boiling point high, about 125°C.

n-Butyl Alcohol

n-Butanol, colourless with a characteristic pungent and irritating odour. Although a good solvent for many plastics its main use in palaeontology is to 'de-water' small waterlogged specimens. It is immiscible with water and displaces it from a porous structure. Inflammable. High boiling point approximately 116°C. In spite of its pungent and irritating odour it is not considered to be a dangerous solvent.

Carbon Tetrachloride

Colourless, heavy liquid with a characteristic pleasant sweet odour. Immiscible with water. Used in palaeontology to dissolve greases, oils or waxes used as separators in casting. Is also used to separate small fossils from a finely divided matrix by flotation. It is non-inflammable but if the vapour comes in contact with a flame or a redhot surface the poisonous gas phosgene is formed and for this reason one should never smoke when it is being used. Its boiling point is low, about 76°C. Like all chlorinated solvents it should be used whenever possible in a fume cupboard and never in a badly ventilated room.

Chloroform

Colourless heavy liquid with a pleasant odour. Immiscible with water. Good solvent for polymethylmethacrylate. Used to weld this material to itself and to dissolve it to make a fast-drying adhesive for the repair of small fossils. Also a powerful grease solvent. Non-inflammable, boiling point low, about 61°C. Its anaesthetic properties are well known so as little as possible should be inhaled and working quantities kept to a minimum. A chlorinated solvent, so never use in badly ventilated places.

Diacetone Alcohol

Colourless, but will darken with age. Miscible with water. Used as a high boiling solvent to prevent 'blushing' in plastic coatings. May be used in 1:10 to 1:20 mixture with acetone to slow the evaporation of solutions of cellulose acetate. Mixed with more volatile alcohols may be used in the solution of other plastics, polyvinyl acetate, etc. Boiling point high, about 167°C.

Dibutylphthalate

Colourless oily liquid. Only used in palaeontology as a plasticiser for polyvinyl chloride. Boiling point above 300°C. Hands should be washed after use to avoid ingestion.

Ether

Diethyl ether, colourless with a characteristic odour. Miscible with water to a slight degree. Excellent solvent for fats, oils and greases. Highly voltatile. Both liquid and vapour are highly inflammable. The vapour can be ignited with explosive violence at below red heat. Has few uses in palaeontology and should be avoided as it can normally be replaced by a less dangerous solvent. Boiling point very low, about 35°C.

Ethyl Alcohol

Ethanol, absolute alcohol, spirits of wine. Colourless with a distinctive vinous odour. Miscible with water. In Great Britain its sale and use is strictly controlled by Customs and Excise regulations. It is hardly ever necessary to use it in palaeontology. There may be some palaeobotanical techniques in which it must be used but as a rule the cheaper industrial methylated spirit may be used in its place. Boiling point about 78°C. Inflammable.

Ethyl Acetate

Colourless with a pleasant smell, said by some to resemble pineapples. Partially soluble in water. Is a good solvent for polystyrene, polymethylmethacrylate and other plastics. Normally used in making adhesives for the repair of fossils when rapid drying is needed. A thin solution of polystyrene in this solvent was the first to be used as a coating in the acid preparation of fossil vertebrates. Inflammable. Boiling point low, about 77°C.

Formic Acid

Colourless with a pungent odour said to resemble that of crushed red ants. Miscible with water. Used as a 10 per cent or less (v/v) solution in water to develop fossil vertebrates. Corrosive to the skin and less safe than acetic acid. Boiling point depends on concentration and is of no interest to the palaeontologist.

Industrial Methylated Spirit

I.M.S. colourless with alcoholic odour. Is mainly composed of ethyl alcohol, the rest being made up of water and a small percentage of methyl alcohol. Miscible with water. Sale and use in Great Britain subject to Customs and Excise regulations. When not obtainable, common methylated spirit (see below) may be used in most cases. Solvent for shellac, polyvinyl acetate, soluble nylon and other man-made plastics. Inflammable. Long exposure to high concentrations of the vapour is dangerous. Methyl content can be absorbed through the skin.

Iso-propyl Alcohol

Iso-propanol. Colourless with characteristic odour. Miscible with water. The preferred solvent for some of the polyvinyl butyrals (i.e. Butvar B98). Slightly more toxic than ethyl alcohol but dangerous levels of vapour should not be reached in a well ventilated laboratory. The solvent is widely used in industry. Inflammable. Boiling point low, about 82°C.

242 Chemical and Natural Substances

Methyl Alcohol

Methanol, wood alcohol. Colourless with an odour differing slightly from and less pleasant than that of ethyl alcohol. Miscible with water. This is a dangerous solvent, very much more toxic than ethyl alcohol. Fortunately there are few uses for it in palaeontology and it should be avoided. It can cause blindness and death if taken internally and the vapour is also dangerous. Inflammable. Boiling point low, about 64°C.

Methyl Ethyl Ketone

Colourless with a pleasant odour, rather like that of acetone. Miscible with water. Chief use as a solvent for polybutylmethacrylate. It is a powerful grease solvent and should not be used to clean the hands. Inflammable. Boiling point low, about 79°C.

Methylated Spirit

In Great Britain a transparent blue liquid composed of ethyl alcohol, water, methyl alcohol, additives to make it unpalatable and a dye. Can be used in place of industrial methylated spirit as the dye, on drying, is fugitive. May be bought freely from oil mongers, decorator's shops and chemists. Miscible with water. The proportion of methyl to ethyl alcohol is higher in this liquid than in industrial methylated spirit. Inflammable. Boiling point low.

Methylene Chloride

Colourless heavy liquid with characteristic odour. Immiscible with water. A powerful solvent for waxes. This is its main use in palaeontology. It is the basis of some commercial paint strippers and varnish removers. It is a chlorinated solvent so should be used with care. Non-inflammable. Boiling point low, about 40°C.

Morpholine

Colourless with a slightly ammoniacal odour. Miscible with water and hygroscopic. Soluble in alcohols. Used in solution with industrial methylated spirit to arrest pyrites decay in large specimens. It reacts with the iron salts in the decomposition products, converting them to hydroxides. In air it decomposes and deposits ammonium carbonate which also acts as an alkaline bank against further decay. The concentrated liquid is corrosive. Boiling point about 130°C.

Savlon
The trade name of a yellow, pleasant smelling antiseptic liquid manufactured by Imperial Chemical Industries and available from local chemists. Used in the experimental treatment of fossils to arrest pyrites decay.

Silicones
Some of these are colourless oily liquids with no odour, others are greases. Immiscible with water. They are used to store small dry specimens which are liable to decay due to the breakdown of iron pyrites. In casting they are used as separators. Inflammable.

Toluene
Toluol. Colourless with an odour resembling that of benezene. Immiscible with water. May be used in almost every case where benezene is recommended. It is a good solvent for polybutylmethacrylate and polyvinyl acetate. Toxic hazards much less than those of benzene or xylene. Inflammable. Boiling point about 110°C.

Xylene
Xylol. Colourless with odour like that of toluene. Immiscible with water. Traditional solvent for Canada balsam. Has few other uses in palaeontology. Inflammable. Boiling point about 139°C. Toxic.

Organic Solids

Cetrimide
A white powder used in solution in alcohols as an experimental cure for pyrites breakdown. Avoid inhaling as this causes great irritation of the nose and throat.

Cellophane
A transparent film derived from cellulose. Commonly used for wrapping. Used in palaeontology as a cheap method of screening

specimens and bench tops from casting resins. Is not soluble in the more common solvents.

Cellulose Acetate

Occurs as powders, fibres, clear film and sheets and in other forms. Used as a consolidating agent in solution in acetone with a low percentage of diacetone alcohol. Easily formed at low heat and is employed to make mounts for small fossils. The basis of some varnishes and adhesives. Non-inflammable.

Cellulose Nitrate

Occurs in several forms from guncotton to celluloid. Avoid buying guncotton for obvious reasons. If a solid form is needed, celluloid is the safest, but even with this there is a very high risk of fire. It is safer to buy ready-made varnishes if you must use this material, but in every way cellulose acetate is to be preferred. Extremely inflammable.

4-Chloro-m-Cresol

A colourless crystalline solid. The crystals are irregular and large. Used in the experimental treatment of pyrites breakdown. The substance is corrosive and poisonous.

Citric Acid

A white powder, soluble in water, used in the chemical development of both vertebrates and invertebrates. In neither case is it very successful. Also used in small amounts to keep iron salts in solution when washing specimens effected by pyrites breakdown. Not poisonous.

Gelatine

A protein substance extract from the collagen in animal tissue. Available as a powder or in sheet form. The sheet form soaked in water is used as a moulding material for casting fossils. Glue is a less pure form of gelatine.

Gum Tragacanth

A natural gum extracted from the plant *Astragalus gummifer*. Obtainable as a powder or in thin flakes. It is not soluble in water to

Organic solids 245

any extent but swells to a sticky mass. Used to cement microfossils to mounts.

Jute
A natural fibre obtained from certain plants akin to the lime tree. Used in palaeontology as a fine floc as a filler in some gap-filling adhesives. A very short staple floc about 1–2 mm is required.

Nylon Soluble
Most nylons are not soluble in practical solvents. One grade made by I.C.I. is dissolved in a mixture of alcohol and water (approx. 10–1 v/v) or in methanol alone, it is a good consolidant, especially for light coloured fossils. With ageing it may loose its solubility.

Oxalic Acid
Salts of lemon. Colourless crystals soluble in water. Mentioned by some authors as a means of developing both vertebrate and invertebrate fossils. Not very successful in most cases. May be used to remove ink stains from paper. This substance is poisonous.

Paraffin Wax
Translucent wax with a white fracture. Sold according to melting point. Those with a melting point between 40°C–50°C are the most useful. Normal use in palaeontology is to save polyester resins in the transfer technique. Do not use on porous material as it is very difficult to remove. Inflammable.

Phenol
Carbolic acid. White crystals but turns pink owing to water taken up from the air. It has a characteristic odour. Only slightly soluble in water but such a solution is a powerful fungicide and disinfectant. Should be handled with care as both the solid and its solution are corrosive to the skin and very poisonous. Burns from phenol, even if only small in size, should receive prompt medical attention, as apart from the corrosive action there could be physiological complications. It is wise to wear rubber gloves and industrial goggles when handling phenol. Volatile.

Polybutylmethacrylate

The author has never been able to obtain this plastic in the solid form in Great Britain. It is colourless and soluble in a wide range of solvents including toluene, white spirit and ketones, and is available in solutions, mostly in toluene. It is used to exclude water vapour and air from specimens treated for pyrites decay and as a coating for bone undergoing acid development. Inflammable.

Polyethylene Glycols

P.E.G. 'Carbowax'. These substances vary from liquids to white wax-like solids according to molecular weight. P.E.G. 4000 is a solid with many uses in palaeontology, including the support of small specimens during mechanical development. Its melting point is low, about 50°C. It is soluble in water. Inflammable.

Polymethylmethacrylate

'Perspex', 'Lucite'. Available as powder, sheets, rods, tubes and other forms. Colourless, soluble in chloroform, ethyl acetate and other solvents. Ethyl acetate is preferable. The sheet is used to make small mounts and a solution is a good adhesive for small specimens especially during preparation by acetic or formic acid. Inflammable.

Polystyrene

Colourless, available as chips or as powder, or as an expanded foam much used in decorating. Readily soluble in ethyl acetate, it makes a good adhesive. Inflammable.

Polythene

A white amorphous material with a high melting point, but this is often deliberately lowered by mixing it with waxes. Available as chemical ware, tubing, sheets, films and household utensils. Is not effected by hydrofluoric acid or by dilute solutions of mineral or organic acids. Inflammable.

Polyvinyl Acetate

P.V.A. White powder, soluble in toluene, alcohols and other organic solvents. A solution in toluene may be used in place of polybutylmethacrylate for the coating of fossils after treatment for pyrites breakdown. Thin solutions in industrial methylated spirit are used as a consolidant and a thick solution in any fairly fast drying solvent, is an excellent adhesive. Also obtainable as emulsions which are good adhesives and on dilution with water may be used as consolidants.

Polyvinyl Alcohol
P–OH. White powder soluble in water. Used as an adhesive in place of gum tragacanth and as a separator in casting.

Polyvinyl Butyral
'Butvar', 'Mowital'. White powder. There are several grades most of them soluble in alcohols but not in ketones. There is one which is soluble in acetone. The best materials for consolidation.

Polyvinyl Chloride
P.V.C. White powder, also as tubing, sheeting, films and in many other forms. The basis of a large range of synthetic rubbers. The powder mixed with plasticisers, is used as a heat-cured moulding compound. In other formulations it forms hot-melt casting compounds. Also exists as co-polymers with vinyl acetate, some of which are soluble in organic solvents have been used as flexible adhesives by archaeologists.

Salicyclic Acid
Either colourless crystals or white powder. Slightly soluble in water. Used to develop certain small calcareous fossils.

Shellac
Reddish-brown natural resin available as disks of about 5 cm diameter known as button shellac, as flakes and small beads. Normally used as solution in industrial methylated spirit. A ready made solution is used by painters and is known as 'knotting'.

Sodium Alginate
A white powder soluble in distilled or de-ionised water. It reacts with calcium ions to form a jelly-like mass. It is the basis of some casting compounds and an aqueous solution is used as a separator in casting from plaster moulds.

Sorbitol
White, sweet-tasting powder. Soluble in water. Used as an additive to gelatine for mouldings.

Terylene
A very strong synthetic material mostly used as fibres for making cloth, ropes and netting. The netting sold for curtains may be used for sieving. A foil, 'Melinex' is made by I.C.I. In its original form it is transparent, but a mirrorised type is available.

Inorganic Liquids

Ammonium Hydroxide s.g. 0.88
Colourless solution of ammonia gas in water with an extremely pungent and characteristic odour. It is corrosive to the skin and the gas emitted has an immediate and unpleasant effect on the nose and respiratory organs if inhaled suddenly. Exposure to high concentrations can be fatal. A strong alkali.

Hydrochloric Acid
Muriatic acid, Spirits of salts. This substance is a gas and the acid is a strong solution in water in which it is highly soluble. Pungent odour. The pure acid solutions are colourless, the commercial grades are yellow. On exposure to damp air, white fumes appear above the acid. Highly corrosive. Used in dilute solutions to extract silicified fossils from limestone.

Hydrogen Peroxide
Colourless liquid with a characteristic odour. Sold in 10 to 100 volume concentration. The figures mean that the gaseous content of oxygen is that many more times the volume of the liquid. Corrosive to the skin. Store in a cool place. Will break up some rocks and is widely used by micropalaeontologists.

Hydrofluoric Acid
The acid sold is a solution of the gas in water. Colourless, fuming liquid with a pungent odour. Highly corrosive and very poisonous. Burns caused by it should be regarded as serious injuries and prompt medical attention should be sought as special

treatment is needed. Both the liquid and the gas will attack glass so it must be stored in plastic or gutta percha bottles and used in heavy duty polythene vessels. It is so dangerous that it is folly to use it when some other acid will do and on no account should it be used outside a fume cupboard. The glass of the cupboard will be etched by the vapour and become opaque. The only acid which will dissolve silicified matrices. Forms an insoluble fluoride with calcium salts.

Nitric Acid
Aqua fortis. The pure acid is colourless, the commercial grades are slightly yellow and give off red fumes. Odour distinctive and pungent. Has few uses outside palaeobotanical techniques. Miscible with water. Highly corrosive, causing burns which are slow to heal.

Sulphuric Acid
Oil of vitriol. A heavy, oily, colourless liquid with a pungent odour. It has a great affinity for water. When making solutions the acid must be added to the water a little at a time with constant stirring as much heat is generated. To mix the water into the acid is courting disaster, for at best it could result in bad burns and at the worst in blindness. Highly corrosive. It is used in some palaeobotanical techniques.

Inorganic Solids

Aluminium Oxide
White powder insoluble in water. It is available in several grain sizes and is used as a fine abrasive.

Ammonium Chloride
Sal ammoniac. A white powder, or in a less pure form as slightly greyish lumps. Soluble in water. It is volatile and is used to whiten fossils before photography. After use, the salt should be first brushed off and then the specimen should be washed.

Asbestos
A naturally-occurring mineral fibre noted for its fire resistance.

Available as a powder, fibres, cloths, rigid boards and other fabrications. Some of the boards are slightly flexible. Colour greyish white. Inhalation of the dust is a health hazard, and can cause lethal damage to the lungs.

Barium Chloride

A white powder or small colourless crystals. Soluble in water. The solution must be made in distilled or de-ionised water as most tap waters contain enough dissolved sulphates for a precipitate of barium sulphate to form. Used in the experimental treatment of small fossils suffering pyrites breakdown and as a test for sulphates. The chloride is poisonous.

Barium Hydroxide

Colourless crystals. Used in solution in water for the same purpose in palaeontology as the chloride. Has the advantage that on exposure to air barium carbonate is formed and this provides a reserve of alkali against future attack. The hydroxide is poisonous.

Calcium Hydroxide

Slaked lime. White powder. Only use in palaeontology to neutralise waste hydrofluoric acid. It is the only safe way of doing this as calcium fluoride is insoluble and the soluble fluorides are poisonous. Calcium hydroxide is slightly soluble in water and a solution is commonly known as lime water. Do not confuse this substance with quick lime which is corrosive and reacts violently with water producing heat.

Copper Sulphate

Blue vitriol, engineer's blue stone. Blue crystals in the hydrated form. The dehydrated substance is white and may be used as a test for the presence of water in which it turns blue. Solutions used to deposit a thin film of copper on de-greased iron or steel surfaces. Lines scratched into such a film are easier to see when cutting the metal to shape. Soluble in water. Poisonous.

Kaolin

China clay. A white powder. Occurs as light and heavy kaolin. The latter is most often used.

Potassium Hydroxide

Caustic potash. A white substance obtainable as sticks, flakes or pellets. Hygroscopic. Mentioned in some techniques, but can nearly always be replaced by the cheaper sodium hydroxide. Both the solid and solutions are highly corrosive to the skin.

Sodium Carbonate

Washing soda, soda ash. Colourless crystals or white powder. A weak alkali sometimes added to water to break down clays.

Sodium Hexametaphosphate

Transparent flakes or a white powder. Solutions in water used to break down clays and other matrices. Forms complexes with calcium ions.

Sodium Hydroxide

Caustic soda. White substance occurs in the same forms as potassium hydroxide and is used for the same purposes. Solutions and the solid are corrosive to the skin.

Sodium Hypochlorite

Normally obtained in solution as household and industrial bleaches. Usually slightly yellow liquids smelling of chlorine. Used to break down coals and other matrices. Keep well away from acids as mixture causes the release of chlorine.

Sodium Thiosulphate

'Hypo'. Colourless large crystals of distinctive shape. Very soluble in water. Used to break up permeable matrices by boiling in a saturated solution and then allowing this to crystallise.

Titanium Dioxide

White powder. Used as a dry colour in painting casts and as a filler to produce a dense white substance with polyvinyl chloride and casting resins.

Cold Setting Compounds

Epoxy Resins

Man-made resins used as the basis for adhesives, potting resins and resins for glass fibre fabrication as well as plastic woods and putties. They are activated by a variety of hardeners. They do not shrink on setting and the hardened material is very difficult to break down by chemical means. More expensive than the polyesters. Some people can, often through careless usage, become allergic to some of the hardeners. Inflammable.

Polyester Resins

Usually a mixture of two man-made resins in which part of one is capable of crosslinking with part of the other in the presence of a catalyst and an accelerator. Widely used to make glassfibre casts and the clearer types for the embedding of specimens either entirely as a protection from the atmosphere or partially as in the transfer method of acid development. All ingredients are inflammable. The set resins will break down, but will not dissolve, in chloroform or methylene chloride.

Polysulphides

Some are liquids and form the basis of some cold setting moulding compounds.

Polyurethane

Akin to the polyester resins. It is used in the manufacture of resistant paints and varnishes. In palaeontology it is mainly used as a rigid foam produced by mixing two liquid parts together. One usually contains an isocyanate and is poisonous and can cause skin irritation.

Silicone Silastomers

Thick, usually white, opaque liquids which on the addition of a catalyst solidify to form an elastic mass. Excellent but expensive moulding materials.

Appendix 1
Special Tools and Techniques

A Polythene Pipette

The materials and apparatus required are: some polythene tubing with an internal diameter of approximately 1 cm and with a wall thickness of 2 mm, a 30 cm length of metal tube with an internal diameter of 3.5–4.0 cm (a piece of waste steam pipe will do), a retort stand and a bunsen burner.

The metal tube is clamped in the retort stand at an angle of about 45 degrees, and at a height which places the lower end just above the zone of no combustion of the bunsen flame. The end of the tube should not be in the flame but very close to it. When the burner is lit a stream of hot air will be carried up the tube and will emerge at the top end. A piece of polythene tube about 20 cm long,

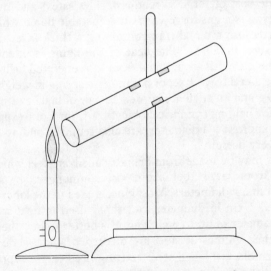

Fig. 27. Apparatus for making a polythene pipette.

is held at both ends parallel to the ground and is rotated close to the top end of the metal tube so that its middle is heated (Fig. 27). When plastic becomes transparent it is removed from the hot air stream and gently pulled. This must be done carefully and a little practice may be needed to get it right, for if it is pulled too sharply the plastic will tear. The hot plastic will stretch into a thin tube connecting the two ends. It must be held in position until it cools and once more becomes opaque or it will distort. Since this takes rather a long time it is best to move the hands from the horizontal to the vertical plane when gravity will help to avoid distortion.

When cold the thin tube is cut in the middle. Now, if the two broad ends of the pipettes so formed are taken in turn and rotated in the hot air stream, the plastic at the ends will melt and form itself into a collar at the top end of the pipette. This helps to keep the rubber bulb in place, which may be fitted as soon as the plastic is cold.

Forging and Tempering Cold Chisels

Before steel can be worked it must be made red hot. For this purpose some sort of forge is required. The safest and most convenient type is a gas forge with an adjustable flame which plays downwards into a metal tray containing a thick layer of 2.5 cm square asbestos blocks. The heat of the flame is intensified by mixing the gas with air pumped from foot operated bellows fixed below the metal tray. A set of hammers, varying in weight from 1 kg to 100 g and an anvil are needed. A pair of blacksmith's tongs, suitable for holding cylindrical objects of from 3 mm (approx: $\frac{1}{8}''$) to 1 cm (approx: $\frac{1}{2}''$) diameter, are also required and an asbestos glove is very useful.

Chisels may be made from cylindrical silver steel which is obtainable from large tool shops or ironmongers in standard lengths of many diameters. Most chisels used in the laboratory do not exceed 1 cm in diameter, but those used in field work may have a diameter of up to 3 cm and it is better to buy these ready made. Other sources of steel are discarded single-ended dental probes, the old fashioned steel knitting needles used for making socks, and the spokes from bicycle wheels.

Special Tools and Techniques   255

It is easier to control the movement of the steel rod during forging, if it is held in the hand, therefore it is better to make two chisels at a time by forging the required shapes at each end of a single rod. The asbestos glove is used for this purpose. Should the end being held become uncomfortably hot, even for a gloved hand, it may be wrapped in a piece of wet cloth.

The steel rod is heated to redness at one end for a distance slightly less than that required, from the cutting edge to the point at which the tapering commences, in the finished tool. It is removed from the flame and the heated end placed flat on the anvil and rhythmically hammered. During the hammering the rod is slowly drawn towards the operator, i.e. the moving steel is passing between the faces of the anvil and the falling hammer. When the metal cools to below red heat the hammering must stop and the rod re-heated. By these means one face of the end of the rod is flattened. It is re-heated and the opposite face treated in the same way. Now the end of the rod has two flat and two half round faces; further re-heating and hammering of the two half round faces in turn produces a rod with a square end.

The heating and hammering are continued until the tool is of the required shape, either a flattened wedge or a point formed by the tapering and converging of the four faces. The shapes are achieved by the 'drawing through' action and are governed by the amount of hammering done on each face. The wedge shape is arrived at by hammering two opposite faces more often than those adjacent to them, and the point is formed by hammering each of the four faces equally. When one end of the rod has been fashioned, it is reversed and the same operation carried out on the other end, then the two tools are separated by sawing through the middle of the rod. The cutting edges or points are finally formed and sharpened by grinding. A broken tool may be repaired using the method described, but as a rule it will be too short to be held in the hand when hot, and tongs must be used. Their use is made easier if a metal ring is available which can be pushed over the handles and hold them together. This maintains a constant grip on the chisel and lessens the strain on the worker's wrist.

When steel is heated and allowed to cool slowly in air it loses its temper and becomes too soft to be used as a cutting tool. A chisel can be tempered to the correct hardness by the following simple method. The cutting end is heated to redness to about 5 cm up the shaft of the tool. It is taken from the flame and allowed to cool to a cherry red and then plunged quickly into a vessel of cold water, large enough for the whole tool to be immersed. When the metal is cold it is dried and one face, for about 5 cm from the cutting edge, is polished with emery paper or a fine cut file. A point about 5 cm from the sharp end of the tool is held in the tip of a Bunsen flame; after a short while it will be noticed that colours begin to appear in the polished steel, at the point at which it is heated, and start to move from the flame towards the cutting edge. The colour furthest from the flame is light brownish yellow, usually referred to as 'straw' colour, and this is followed by a series of purples and blues. The moving band of colour is watched carefully and just before the 'straw' colour disappears from the tip, the whole tool is rapidly immersed in cold water and left to cool down.

The final stage is to sharpen the edge or point; this is best done on a hand-operated grind stone, because the steel must not be heated by friction, and this type of grinding wheel is more easily controlled than one operated by a motor.

The making and tempering of a perfect chisel takes a little practice but a usable tool is normally produced at the first attempt. Preparators very often need tools of a shape which cannot be bought and for this reason it is worthwhile extending one's skill in tool making to encompass objects more complicated than chisels.

Two Pieces of Apparatus for use with Polyethylene Glycol 4000

When working on small fossils of any type it is often convenient to have a few cubic centimetres of polyethylene glycol 4000 constantly available in a molten state and to have an easy means of softening small areas of the solid wax. Two pieces of apparatus, which have to be made by the preparator himself, have been devised to meet these needs. The first is a melting pot which can be in constant use with little or no danger of fire.

Two 35 mm film cans are needed; the bottom is cut off one at about 3 mm above the base and the bottom of the other at 2.0 cm above the base. A hole equal to the external diameter of the can is drilled through the centre of a piece of flexible asbestos board 8 cm square. The longer bottomless can is passed through the hole and the lower end is slightly widened by pushing it onto a cylindrical piece of wood whose diameter is slightly greater than that of the can. The other shortened can is then pushed bottom upwards into the expanded base of the tube and if it can be pushed approximately 3 mm in it is withdrawn and 'Araldite' applied to the contact surfaces and the whole re-assembled. We now have a vessel from whose base hangs a 2 cm skirt of aluminium.

A second piece of asbestos about 11.5 cm square is now needed; to the four corners of one surface small corks are attached. On the other side a holder which will take a 6 V, 18 W car headlamp bulb or a microscope lamp bulb of the same rating is fitted. The holder must be made either of ceramic or metal; plastic will not do. If a suitable holder cannot be purchased, Fig. 28 shows how one may be made to take a bulb which has two metal contacts in its base. The materials required are two strips of thin aluminium sheet which are bent to make a collar for the bulb, two pieces of flat spring brass, such as can be removed from an exhausted 4.5V torch battery, two small nuts and bolts and some pieces of flexible asbestos board or a heat resistant laminated plastic, such as paxoline and some 'Araldite'. The exercise of a small amount of ingenuity will suggest ways of making a holder for other types of bulb.

The bulb holder is fixed to the 11.5 cm square of asbestos so that the centre of the lamp, when in the socket, coincides with the

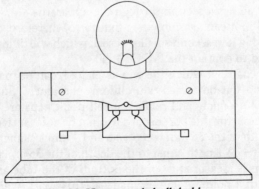

Fig. 28. Home-made bulb holder.

centre of the board. The socket is fitted with a length of 5A flex and connected to a 6V microscope lamp transformer, which must be rated for use with an 18 W bulb. This is important, for a transformer intended for lower wattage bulbs would certainly be ruined and might cause a fire. A radio equipment valve heater transformer capable of reducing the local mains voltage to 6.3 V and of delivering a current of at least 4 A may be used, but if it is it must be housed in a suitably insulated or earthed container. If you have little or no knowledge of electricity the only safe course is to have this made by a qualified electrician.

The board through which the can is passed is now clamped in a retort stand and placed above the lamp. The bottom of the can should not be in contact with the top of the lamp, which should be in the centre of the aluminium skirt. A thermometer is placed in the can and the height of the base in relation to the top of the lamp adjusted to the point at which the temperature rises most rapidly.

The electricity is switched off and the distance between the top of the lower board and the bottom of the upper is measured. Four pieces of asbestos board are cut to this measurement and to widths from which four sides of a box can be built up to contain the can. This may be done by glueing the pieces to the underside of the top board and to each other. Ventilation holes are drilled in each side piece, level with the bulb and another row below this. The base may be fitted to the box with four small metal lugs cemented to the sides and bolted through the base. 'Araldite' or an equivalent epoxy adhesive is used in the assembly of the unit. Figure 29 shows a section through the apparatus.

The 35 mm can assembly is a fixture so, therefore, the wax is not melted in it; a second smaller vessel, which will fit into the can, is required to contain the P.E.G. 4000.

A 'saucepan', in which the wax is melted, can be made from a cylindrical, flat-bottomed aluminium container. Some pharmaceutical products and refills for ball point pens are packed in tubes suitable for this purpose; they must be at least twice as long as the depth of the 35 mm receptacle and 4 mm less than its internal diameter. A length equal to the depth of the 35 mm receptacle plus 6 mm is measured from the bottom of the tube and a mark made at this point. A cut is made at this mark which passes through all but a 6 mm arc of the circumference of the tube. From

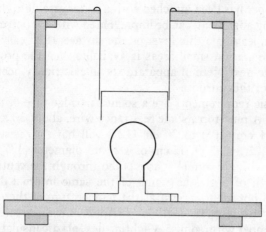

Fig. 29. Section through polyethylene glycol 4000 melting apparatus.

the ends of this arc of solid metal, parallel lines are drawn upwards to the mouth of the tube and cuts are made down their entire length. The greater part of the walls of tube above the transverse cut will then fall away leaving a small pot with a 6 mm wide strip of metal attached to its rim; this is used to make the handle of the saucepan. The length of the projecting 6 mm strip is reduced to approximately 7.5 cm and it is then bent downwards to form an angle of about 100° to the side of the pot. Strips of asbestos board 6 mm wide and 7.0 cm long are cemented to the flat sides of the handle with 'Araldite'. A small spout to facilitate pouring of the molten wax is pinched into the edge of the 'saucepan.' In use, the 'saucepan' is filled with polyethylene glycol 4000 and placed inside the 35 mm can. The apparatus is switched on and the wax melts rapidly. The 'saucepan' may remain in the heating chamber for as many hours as required, because the apparatus is designed to make it impossible for the wax to overheat.

A Small Heating Tool

The rapid attachment of a small fossil to a mount for examination or development, and its equally speedy removal is often desirable. Such mounts can be made from pieces of hardboard to which a piece of paper has been attached with a water insoluble glue. The paper, being absorbent, can be impregnated with hot polyethylene glycol wax, leaving a thin layer on the surface. If an easy way of melting the wax in small areas is available then the problem is solved. The next piece of apparatus is an electrically heated tool designed for this purpose.

The basic requirements are a spent cartridge fuse of the type fitted in 13A plug tops, some resistance wire, about 30 s.w.g., of which 25.4 cm (10″) to 30 cm (12″) will have a resistance of between 4.25 and 5 Ω, 13 cm of 1.27 cm diameter ($\frac{1}{2}$″) wooden dowelling with a 6 mm ($\frac{1}{4}$″) hole bored through the centre or an equal length of a plastic tubing of the same internal diameter which will not melt or burn at the temperature of the element, some 5A fuse wire, a length of 18 to 20 s.w.g. double cotton covered copper wire, some 5 A lighting flex, plastic insulating tape and a piece of polyvinyl chloride tubing with an internal diameter of 1.27 cm (0.5″) or slightly less. A 6 V, 6 W bulb is absolutely necessary.

Nichrome is a suitable resistance wire and tables are published giving its resistance in ohms per yard (0.9 m approx). Successful elements have been made from wire taken from 500 W and 1000 W spirals sold for replacements in electric fires. If this type of wire is used it is easier if some means of measuring low resistance is available, otherwise the correct length must be found by experiment; of course once a length of wire is cut from a replacement element it is rendered useless for its original purpose and should on no account to be used to repair an electric fire.

The cartridge fuse forms the body of the element and is prepared by heating the metal caps at each end and removing them. The granular filling of the body is pushed out with a thin piece of wire. We now have a ceramic tube.

The correct length of resistance wire having been established, this length plus 1.27 cm ($\frac{1}{2}$″) is cut and straightened by holding each end in a pair of pliers and pulling. At about 3 mm ($\frac{1}{8}$″) from each end the ends of two pieces of 5 A fuse wire are bound in a tight spiral of about four turns round the resistance wire. The ends of the resistance wire are bent over these spirals and squeezed tightly onto them with a pair of pliers. Solder is then applied to the junction of the wires. There needs to be about 7.5

cm of the fuse wire extending from each joint.

The resistance wire is now wound into a spiral on the smooth shank of an 11/64″ (4.29 mm approx) twist drill. This is done because the wire is springy and if it were wound directly onto the ceramic tube a tight fit would not be achieved. The spiral is then removed from the drill shank and transferred to the ceramic tube and the turns spaced so that no one is in contact with its neighbour (Fig. 30a). The fuse wires are wound tightly round the

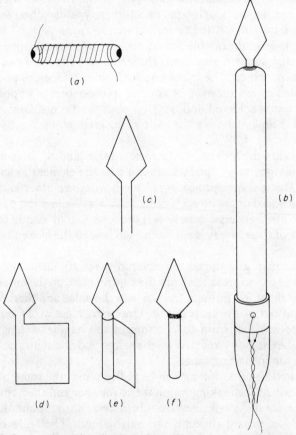

Fig. 30. Low voltage heating iron for polyethylene glycol 4000. (a) Element coil; (b) Complete heating iron; (c) Bit made from thick metal; (d), (e), (f) Three stages in making a bit from thin metal.

tube at each end for at least two turns and the turns soldered together. This is not absolutely necessary but makes the element less liable to damage in use.

A strip of asbestos paper, as wide as the tube is long, is wrapped tightly over the resistance coil leaving the copper wires unenclosed. After two layers have been wound on, one of the copper wires is bent to lay on the asbestos; care should be taken to see that it is widely spaced from the similar wire at the other end. Two more layers of asbestos paper are wound over this wire and is then secured with sellotape, or another type of adhesive tape which will not char at the operating temperature.

Two pieces of the double cotton-covered wire are stripped of insulation for about 6 mm from the ends, and the bare ends are tinned with solder. The wires leading from the asbestos package are twisted round these ends as close to the asbestos as possible and the joints soldered and covered with plastic insulating tape and just below the tape the cotton covered wires are twisted together.

The rigid tube for the handle is now taken and the hole at one end is enlarged to a depth which will take the element assembly, leaving about 2 mm projecting. 1.0 cm from the other end two holes are bored opposite each other which will allow the passage of the cotton covered wires. A groove equal in depth to the diameter of these wires is cut from each hole to the free end of the tube.

The element and the cotton covered wires are placed into the handle and the wires taken from the central hole, up the outside of the tube, threaded through the holes in the sides and then pulled down and out of the central hole. One wire is cut at 2.5 cm and the other at 4 cm from the bottom of the handle. A length of lighting flex is now soldered to the wires and the joints covered with plastic insulating tape.

Polyvinyl chloride tubing can be pulled out in the same way as was described for making a polythene pipette, and the same apparatus is used. A conical cap of this material is made and the 5 A lighting flex passed through the narrow end. The wide end is pushed over the end of the handle to enclose about 2.54 cm. The finished tool is shown in Fig. 30b.

Bits for this instrument may be made from copper wire which

Special Tools and Techniques 263

will pass into the ceramic tube; one end of the wire is flattened by hammering. They may also be made from aluminium sheet; for thicker gauges the shape must be cut out and the tang formed to fit the tube by hammering and filing (Fig. 30c). Figure 30d,e,f show how bits may be made from metal sheet about 1.5 mm thick.

This tool must not be connected directly to the 6 V transformer but must be wired in series with a 6 V, 6 W lamp. If this is not done the tool will be destroyed and the transformer damaged. Before putting a newly made tool into use it must be subjected to a safety test.

It should be clamped in a retort stand above a tray of sand, connected through the 6 V, 6 W lamp to the transformer and left to

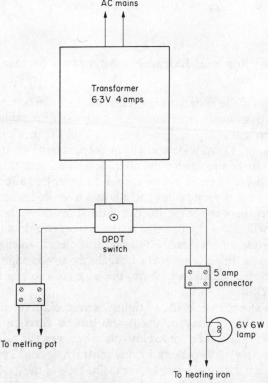

Fig. 31. Suggested circuit for the operation of the melting pot and heating iron from the same transformer.

run for at least an hour. It should be under constant watch during this time. The handle in the region of the element should then be checked for signs of charring and carefully touched to establish that it is not too hot for comfortable handling.

If you make these two pieces of apparatus it is vital to remember that the transformer used must be of the right rating and that in the case of the electrically heated tool it must not be connected directly to the 6 V supply but must go through a 6 V, 6 W lamp. The lamp is there to limit the current to 1 A even if the tool goes short circuit. If a variable transformer is available the heat of both instruments may be lowered by lowering the voltage. Figure 31 shows an arrangement by which both pieces of apparatus may be operated from the same transformer, without constantly connecting and disconnecting wires.

A Glove Box and Extraction Apparatus for use with the 'Airbrasive'

The basis of the apparatus is a well made wooden box; the method of construction is not important and the dimensions may be varied to suit individual needs. A box with the inside dimensions 58 cm long, 43 cm wide and 18 cm deep, is large enough to cater for the size of specimen most often developed with this machine.

Two holes 12 cm in diameter with their centres 32 cm apart are cut in one of the long sides and a hole 5.6 cm in diameter is made in the centre of the other. In both of the short sides a hole is drilled which will allow the passage of the 'Airbrasive' handpiece. The box is fitted with a glazed lid having side pieces which will fit over the outside dimensions of the box. On the inside edge of the lid a foam rubber gasket, which fits the edges of the box is attached (Fig. 32).

Two sleeves are made of tightly woven cloth (or a pair of industrial sleeves may be used) and passed through the 12 cm diameter holes and clamped into place between the side of the box with two plywood gaskets, having central holes of 12 cm diameter and external diameters of 14.5 cm. The tops of the gaskets are cut off to accommodate the overhanging edge of the lid and are fixed with screws (Fig. 33).

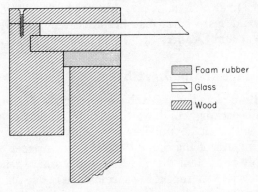

Fig. 32. Section through part of the lid of glove box to show construction.

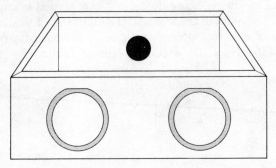

Fig. 33. Glove box without lid.

To avoid damage to the specimen if it is dropped, the bottom of the box is lined with a pad of 1 cm thick foam rubber sealed into a polythene bag; the polythene cover makes the pad easy to clean.

The suction required is provided by a vacuum cleaner. It must be of a type in which coarse particles from the 'Airbrasive' cannot come into contact with the motor. An excellent machine for this purpose is the 'Nilfisk' Model G.A.70, fitted with a microfilter. The 'Nilfisk' runs very quietly and there is a choice of two speeds. The makers supply spare parts, so it is very readily adapted for this apparatus. A female socket, such as that attached to the cleaner can be fitted to the 5.6 cm diameter hole at the back of the box and a male plug, to fit into this, can be fitted to the hose. The cleaner and spare parts may be purchased from Messrs Tellus Super Vacuum Cleaners Ltd.

The Storage of Small Teeth

The bottom of a glass test-tube or small glass vial is covered with a resiliant packing; cut down toilet tissues will do, but not cotton wool. A slit about 1.5 mm deep is cut into that end of a cork which fits the mouth of the tube. A strip of thick paper or thin card, preferably of a colour that is fast to water and to general solvents and contrasts with that of the specimen, is cut so that it is slightly less wide than the diameter of the tube and about half its length. One end of the card is stuck into the slit in the cork with polyvinyl acetate emulsion; the other end is cut to an angle of about 90°, the point of the angle being central. The specimen is stuck by the roots to the point of the angle. The adhesive used should be readily soluble in water, so that should removal be necessary it can be done with ease. If the specimen falls off in the tube its absence

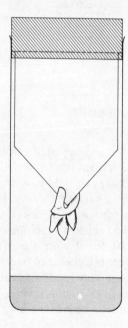

Fig. 34. Storage tube for small teeth.

from the card can be seen readily and it will be caught on the padding at the bottom of the tube. The width of card reduces the possibility of the specimen being damaged when it is either being removed or replaced in the tube. An all round view of the tooth is thus available (Fig. 34).

Mounts for the Storage of Small Jaws and Similar Specimens

In Chapter 5 it was said that small jaws and other specimens, which have been completely freed of matrix, are best stored in transparent plastic boxes on specially constructed mounts. Boxes of this type are made with lids that are hinged, are separate and lift off, or are separate and slide in and out in grooves cut in two opposite sides. The third variety is the safest for the present purpose. The support is made from strips of polymethylmethacrylate or polyvinyl acetate sheet about 1–1.5 mm thick.

The main requirements are that the mount should fit into the box so that, when the lid is closed, there is the minimum clearance between it and the six sides of the box, thus restricting its movement in any direction to a fraction of a millimetre, yet when the lid is removed making it possible for the mount to be taken from the box and replaced, without the exertion of any force. Further, the support must be designed to give an unimpeded view of all aspects of the specimen. Such a structure can be fashioned from two pieces of plastic strip, each bent to form three sides of a rectangle, one made to conform closely to the inner contours of three sides of the box and the other to fit inside the first.

For the sake of clarity, let us assume that the mount is to be fitted into a perspex box whose internal dimensions are 5.0 cm × 3.5 cm having a depth of 1.5 cm measured from the bottom to the base of the grooves into which the lid slides. A strip of plastic sheet 1.5 cm wide and approximately 22 cm long is cut and, let us assume that it is 1.5 mm thick. The thickness of the strip must be taken into account when calculating the lengths required to make the two components of the mount. For the outer member, the length of a long side of the box plus twice the length of a shorter side minus twice the thickness of the strip is needed, that is: 5.0 cm + 7.0 cm − 0.3 cm = 11.7 cm. From each end 3.5 cm is measured off and these points marked, using a small metal try square and a scriber. In turn, the plastic is heated along these lines, by the flame described in Chapter 8, and bent at right angles. The simplest method for making the bends at the right place is to lay a wooden ruler along the scribed line in the softened plastic, hold it in both hands with the thumbs pressing the unheated area against the back of the ruler, place the short arm of the strip onto a smooth block of wood, press down lightly and then turn the ruler onto its edge until it forms a right angle with the block. It must be held in this position until the plastic cools and becomes rigid. When both angles have been made the frame is tested to establish that it fits into the box and can be taken out without difficulty; should adjustment be necessary one angle is reheated, the strip straightened and the bend made at a point which eliminates the error. If either of the 3.5 cm sides proves to be too long the excess may be removed with glass paper. It is important that the fit should conform to the specifications previously stated and it may take a beginner some time to get it right; with practice adjustment is seldom necessary. A satisfactory fit having been achieved, any smoke deposit left by the flame is washed off with detergent and a small area in the 5.0 cm side is roughened with glass paper to provide a surface on which the registered number of the specimen is written.

The second component is required to fit inside the first and to have short arms 1.5 cm long. By referring back to the calculations used to find the length of plastic needed to make the first component it will be seen that its internal length is 4.7 cm, so now a strip 4.7 cm + 2(1.5 cm) − 0.3 cm = 7.4 cm is required. Distances of 1.5 cm are marked from each end and the strip bent at these points as previously described. A slit approximately three quarters of the length of the specimen to be mounted is cut in the middle of the long side and parallel to its edges. This frame is fitted into the first component so that the ends of the 1.5 cm arms rest upon its base and their outer surfaces are in contact with the inner sides of the 3.5 cm arms. This places the upper surface of the second component at approximately 1.5 cm above the base of the first. The two pieces are now chemically welded as described in Chapter 8. When the welds have hardened, the mount is placed in the box and it is established whether or not the lid will slide into position;

should it be fouled the mount is placed on its side on a sheet of 00 glass paper and rubbed down until it will allow closure.

It now remains to attach the specimen to the mount. This is always a delicate operation, but it is made easier if all the tools and materials are arranged in the order in which they will be needed on a large sheet of white paper. The mount should be stood upon its base, a steel rule passed between the two sections and heavy weights placed on each end, so holding the plastic assembly immovable. The slit in the inner member is filled with molten polyethylene glycol 4000 and the specimen is set into it; in the case of jaws, with the teeth upwards. Vacuum tweezers are the best tools for lifting and positioning the specimen, because as soon as it is correctly placed and the vacuum is released, cold air is drawn over the wax which sets immediately. In the absence of this tool the specimens are best handled with a pair of swan-necked forceps, whose tips have been shod with polythene capillary tube, however, there is the difficulty that the specimen has to be held steady until the wax has solidified, it therefore helps if an assistant is available who, at the given word, will blow air across the wax with a rubber blow ball.

The advantage of this type of mount should now be apparent. When the specimen, on its support, is placed in the box and the lid closed, it is protected on all sides and because of the excellent optical properties of perspex, superficial examination may be made without removing it from the container. Since the mount is capable of only slight movement, the fossil is unlikely to be damaged if the box is dropped. Microscopic examination cannot be carried out while the specimen is in the box; the mount may be taken out without difficulty and the specimen need not be handled, it can be placed in any position desired and every part, except the small area embedded in the wax may be studied. No matter in what position it is placed on a bench top the specimen cannot come into contact with the surface (Fig. 35).

There is one possible danger in the microscopic examination and photographing of specimens mounted in this way. Both processes require concentrated beams of light from powerful lamps to be focused onto the fossil and such beams carry considerable heat so the polyethylene glycol wax, which has a low melting point, could liquefy causing the specimen to fall off the

Appendix 1

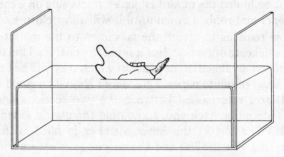

Fig. 35. Plastic frame for storage of small jaws.

mount. This can be prevented by placing heat sinks made of quarter inch plate glass (approx 6 mm) between the light source and the specimen. These must be strongly supported, for glass of this thickness is very heavy and should it fall a unique specimen could be destroyed. In the unlikely event of it being absolutely necessary to remove the specimen from the mount, the properties of polyethylene glycol 4000 make this safe and easy.

Fibreglass Trays for use in Chemical Development

Fibreglass trays of any shape can be built up in properly constructed cardboard moulds. The easiest trays to make are those which are either square or rectangular. The cardboard does not need to be very thick, 3 mm or less has been used successfully. There is an advantage in using glazed card, but it is not essential.

To construct a tray 25 cm square with sides 2.5 cm deep, a piece of cardboard 30.5 cm square is needed. A pencil line 2.5 cm from each edge is drawn for the full length of the card. Using a steel straight edge and a sharp knife the cardboard is cut half way through its thickness for the full length of the pencil line and the 2.5 cm squares formed at each corner are cut out neatly. The sides can now be folded to form a cardboard tray roughly 25 cm square with 2.5 cm deep sides. The corners of the tray are held in place with adhesive paper tape applied to the outside. The whole is now given several coats of a thick solution of shellac and allowed to dry; then the inside of the tray is given several applications of a wax polish; ordinary floor polish does very well. A gel coat of polyester resin is now applied to the inside of the tray; if Trylon GC 150 P.A. is used this needs to be mixed with 2 per cent of liquid catalyst. When required, another gel coat may be applied to the first after it has hardened.

As many strips of chopped strand glass mat, just over 30.5 cm long and about 5 cm wide, as are required to cover the bottom of the tray twice are cut. A strip of mat is placed in the mould so that one long edge is up against a side and the ends of the strip are just above the edges of the mould. Some polyester resin is mixed and stippled thoroughly into the glass using a stiff brush. Trylon WR 180 mixed with 2 per cent accelerator, 2 per cent catalyst and 15 per cent Prefil F is one resin which may be used. It is important that the mat be thoroughly impregnated with the resin and that it be pushed into contact with all surfaces of the mould which it covers. The next strip is applied in the same way and slightly overlaps the first and the process is continued until the length and two sides of the mould are covered. It is easier if this layer is allowed to set before the next is applied. The second layer of strips is applied at right angles to and overlaying the first; this will cover the remaining two sides of the mould and give a bottom to the cast which is two layers of glass mat thick. It may be that the corners of the mould have not been completely covered but this can be rectified by applying small pieces of resin-soaked glass mat to the voids.

The tray should be left to harden for two days when the rough edges can be cut off using a hack saw blade, or a diamond dental disc. The trimming should be done under a dust extractor or, if this is not available, a mask should be worn and the hands and forearms protected with gloves and sleeves as the dust can be an irritant to the skin.

When the trimming is finished the cardboard may be removed by immersion in hot water and holes for drainage bored in the bottom of the fibre glass tray.

The technique of using cardboard moulds can be applied to fairly complicated shapes. The application is only limited by the flexibility of the cardboard and the ingenuity of the person using the method.

Appendix 1
Method used in Shaping Expanded Polystyrene

Expanded polystyrene, in its normal state, is highly inflammable but it can be treated to render it either fireproof or fire resistant. The plastic, in the latter form, is on general sale, but it must be ordered specifically or there is a danger that you will receive the cheaper inflammable variety. It is widely used in interior decorating and as an insulator in the building trade and can be obtained as tiles, sheets or blocks from merchants supplying these trades.

The material is easy to work; it can be carved with very sharp tools, cut with fret saws and, owing to its relatively low melting point, can be fashioned with hot tools. The use of the latter has the advantage that it creates no dust; since dust resulting from working with cold tools is electrostatic and clings to the hair and clothing, this is not an insignificant factor.

A battery-operated tool consisting of a resistance wire held taut between two arms, resembling those of a fret-saw frame, is sold by builder's merchants and craft shops. This is capable of cutting out very complicated shapes in the expanded plastic and may be used when making hollows to contain irregular shaped fossils. The substitution of a low voltage transformer for the battery is advisable because the current drawn is quite considerable and batteries soon become exhausted and when this type of work is not called for very often, they can deteriorate in storage. Where a commercially produced tool of this type is not obtainable it is not difficult to make one, provided that the preparator has a knowledge of the basic laws of electricity.

A solder gun can be modified to carve larger pieces of expanded polystyrene, but it must be a type with the bit formed of a piece of wire which has the same diameter over its entire length. This can be fashioned into a loop and will slice out pieces of the plastic, much as a gouge carves wood. The heat can be controlled, after a little practice, by constantly switching the gun on and off. The heating iron, described elsewhere in this appendix for use with polyethylene glycol wax, can be modified for use with expanded polystyrene by changing the resistance of the element, but if this is done care must be taken to avoid overloading the transformer. For the transformer suggested, the resistance of the element must not be lower than 2Ω and it will be necessary to check by experiment that the material used for the handle will withstand the extra heat. Low voltage soldering irons can be used for working in expanded polystyrene, provided that the voltage applied is regulated to produce only sufficient heat to cut the plastic without charring or melting it to a point where it sticks to the iron and threads are drawn off when the tool is lifted. It is not possible to give details as to how this may be achieved, because the wattage rating of these tools varies, but in general where a variable transformer is not available, the voltage applied to the iron may be dropped by connecting 6 V car headlamp bulbs in series with the tool. Resistors employed in electronics may be used in place of lamps, but it is vital that their wattage rating should be higher than the actual energy dissipated in them, otherwise they will burn out or become very hot.

Storage mounts, of course, must be designed to fulfil the needs of individual specimens and this occasionally requires that two or more pieces of expanded polystyrene be stuck together. The plastic is very soluble in the solvents on which the majority of commercial adhesives are based, therefore only those which are made specifically for this purpose are suitable. Polyvinyl acetate emulsion is an excellent adhesive for joining the plastic in the expanded form.

Although most commonly used for the protection of fragile specimens in storage, mounts for heavy but delicate skulls have been made from large blocks of the material. On rare occasions a block carved to fit the contours of the palate and base of the skull of animals as large as *Dinotherium*, has been used to protect these regions during the preparation of the roof of the skull.

A Useful Small Tool for Moulding

Emphasis has been put on the need for a very sharp and neat finish to the edge of an embedding material against a fossil to be moulded. Small spatulas can be obtained but none small enough to make working to the ornamented edge of a specimen easy. A tool made from 12 s.w.g. Nichrome wire has been found to be very useful.

One end of approximately 13 cm of the wire is gripped in a vice and the other by a pair of pliers; it is pulled and will straighten. A helix of tinned copper wire, 30 A fuse wire will do, is tightly wound from about 2.5 cm from one end to within about 2.5 cm of the other. Solder is applied to the copper wire to convert the helix into a cylinder. The ends of the wire are then beaten flat on an anvil or smooth piece of iron with a small hammer and then shaped either to a point or a curve on the coarse side of an axe stone or an equivalent hone.

Using a pair of round-nosed pliers the ends are given a slight upward curvature. The under surface of the curves can now be polished and slightly rounded on a fine hone.

Nichrome wire can be worked cold. The copper wire spiral and the solder stiffen the body of the tool and provide a good grip.

Appendix 2
Materials, Manufacturers and Suppliers

This appendix is in the form of an alphabetical list of articles with the names of the suppliers in the British Isles and, where it is thought that it may be helpful, the manufacturers. The full address of the firms mentioned are given at the end. It has been arranged this way to make reference easy.

Most of the firms named issue catalogues or lists of the articles they stock and all have many more than are mentioned here and these catalogues are essential to a laboratory.

There are, of course, other suppliers of some of the items listed and these can be found from trade directories, the 'yellow pages' of local telephone directories and through advertisements in newspapers or special publications such as 'Exchange and Mart'.

ACETIC ACID (in bulk)
   Victor Blagden & Co. Ltd.
AEROSOL SILICA (see 'Santocel')
AIRBRASIVES S.S. WHITE (Also abrasive and spare parts for same)
   G.E.C.-Elliott Mechanical Automation Ltd.
A. J. K. DOUGH
   Frank W. Joel, Museum Laboratory & Archaeological Supplies Ltd.
ALUMINIUM SHEET (Thin, approximately 0.25 mm)
   J. Smith & Son (Clerkenwell) Ltd.
   Sometimes obtainable from local ironmongers.
BEDACRYL 122X
   Manufacturer: Imperial Chemical Industries Ltd.
   Supplier: Frank W. Joel, Museum Laboratory & Archaeological Suppliers Ltd.
BRUSHES, ARTISTS
   Any good dealer in artist supplies.
BRUSHES, GENERAL PURPOSE, BRIDLED GLUE, FITCHES, ETC.
   A. Leete & Co. Ltd.
   Enquire about minimum order. Ask for brushes with unvarnished handles.

Appendix 2

BRUSHES, RESIN
  A. Tiranti Ltd.
  Trylon Ltd.
BUTVAR B98
  Frank W. Joel, Museum Laboratory & Archaeological Supplies Ltd.
CARBON DIOXIDE, COMPRESSED GAS
  D.C.L. (CO_2 Division). See suggested sources for local supplies.
CARBOWAX 4000 = POLYETHYLENE GLYCOL 4000
  Carbowax 4000 from Frank W. Joel, Museum Laboratory & Archaeological Supplies Ltd.
  P.E.G. 4000 from Hopkin & Williams Ltd.
'CAVITRON'
  Claudius Ash & Sons Co. Ltd.
CELLULOSE ACETATE SHEET
  There are several manufacturers. May & Baker (Plastics Div.) Ltd make 'Rhodoid'. Small users should try stationers, artist suppliers and hobby shops. Also advertisements in the suggested sources.
CHEMICALS
  A. Gallenkamp & Co. Ltd. (sell B.D.H. products).
  Hopkin & Williams Ltd.
  May & Baker Ltd.
CHEMICAL APPARATUS AND GENERAL LABORATORY REQUIREMENTS
  A. Gallenkamp & Co. Ltd.
CHISELS, COLD
  Large tool shops.
CUPS, PAPER
  Mono Containers Ltd.
DENTAL ENGINES AND STANDS
  Citenco Ltd. These machines are relatively cheap and hard wearing.
DENTAL MALLETS
  Anthrogyr Power Mallet No. 1650 from: Marcel Courtin.
DENTAL TOOLS, BURRS, HANDPIECES, IMPRESSION COMPOUNDS, etc.
  Claudius Ash & Sons Co. Ltd.
  Cottrell & Co.
EPOXY, ADHESIVES AND RESINS
  C.I.B.A. (Araldite).
  Shell Chemicals U.K. Ltd (Epikote).
EMULSIONS, POLYVINYL ACETATE
  Manufacturer: Vinyl Products Ltd.
  Supplier: Frank W. Joel, Museum Laboratory & Archaeological Supplies Ltd.
ENGRAVERS (see Speed Engravers)
'FIBRENYL DOUGH' (see A. J. K. Dough)

Materials, Manufacturers, Suppliers 277

FORCEPS, SMALL
  A. Gallenkamp & Co. Ltd.
  Watkins & Doncaster.
GLASS FIBRE, ALL TYPES REQUIRED FOR CASTING
  Fibre Glass Ltd. A large firm with a very wide range of glassfibre materials. Sell in reasonably small quantities and issue literature on their products.
  Strand Glass Co. Ltd.
  A. Tiranti Ltd.
  Trylon Ltd.
GOGGLES, INDUSTRIAL
  A. Gallenkamp & Co. Ltd.
  Stocked by some large tool shops.
HAMMERS, CLUB
  Tool shops supplying equipment for bricklayers.
HAMMERS, GEOLOGICAL
  Gregory Bottley & Co.
HAMMERS, SCULPTOR'S
  A. Tiranti Ltd.
JUTE FLOC
  Frank W. Joel, Museum Laboratory & Archaeological Supplies Ltd.
'MELINEX', CLEAR FILM
  Manufacturer: Imperial Chemical Industries (Plastics Div.) Ltd.
  No small supplier known, but try Frank W. Joel and Watkins & Doncaster. Some transparent cooking foils made for oven roasting may be employed, after tests have proved that they are not attacked by the material to be used.
'MELINEX', MIRRORISED
  George M. Whitley Ltd.
  Sold in reasonable quantities according to weight and price. Ask for sample swatches of the gauges available. The exhibition potential of this material has not been fully exploited.
'NITROMORS'
  Frank W. Joel, Museum Laboratory & Archaeological Supplies Ltd.
  Paint shops.
NYLON, SOLUBLE
  Manufacturer: Imperial Chemical Industries Ltd.
  Supplier: Frank W. Joel, Museum Laboratory & Archaeological Supplies Ltd.
'PERMABOND'
  Staident Products Ltd.
PINS, STEEL HEADLESS 15 S.W.G.
  Armstrong Cork Co. Ltd.

Appendix 2

PIN VICES
  The 'Eclipse' range from all large tool shops.
  Lighter types: A. Gallenkamp & Co. Ltd.
              Watkins & Doncaster.
PLASTER, CASTING OR FINE BAKED INDUSTRIAL
  Graham (Builder's Merchants) Ltd.
  A. Tiranti & Co.
  Builder's merchants.
PLASTERER'S SCRIM
  A. Tiranti & Co.
PLASTIC MESH, 'TYGAN'
  Fothergill & Harvey Ltd.
PNEUMATIC POWER PEN V.P.2
  Desoutter Bros. Ltd.
POLYBUTYLMETHACRYLATE (see 'Bedacryl' & 'Vinalak')
POLYESTER RESINS
  Manufacturer: Scott Bader Ltd.
  Suppliers: Strand Glass Co. Ltd.
           A. Tiranti Ltd.
           Trylon Ltd.
POLYETHYLENE GLYCOL 4000 (see Carbowax 4000)
POLYMETHYLMETHACRYLATE, COLD SETTING POWDER AND LIQUID
  'Metallurgical Mounting Medium'.
  North Hill Plastics Ltd.
  (Powder 225 gm. Liquid approx. 114 gm)
POLYMETHYLMETHACRYLATE, POWDER (high molecular weight)
  B.D.H
  A. Gallenkamp & Co. Ltd.
POLYMETHYLMETHACRYLATE, PERSPEX SHEET
  Manufacturer: Imperial Chemical Industries (Plastics Div.) Ltd.
  Suppliers: D. & J. Plastics (Croydon) Ltd.
           Visijar Laboratories Ltd.
  Also artists suppliers and hobby shops. Consult suggested sources for local suppliers.
POLYSULPHIDE, COLD SETTING CASTING RUBBER, 'SMOOTH-ON'
  Walter P. Notcutt Ltd.
POLYTHENE, SHEETING AND BAGS
  Transatlantic Plastics Ltd.
  See suggested sources for local suppliers.
POLYTHENE, TUBING
  A. Gallenkamp & Co. Ltd.
POLYVINYL ACETATE, EMULSION (see Emulsions)

POLYVINYL ACETATE, POWDER
  Hopkin & Williams Ltd.
POLYVINYL BUTYRALS (*see also* Butvar B 98)
  Manufacturers: Monsanto Chemicals Ltd.
              Farbwerke Hoechst A.G.
  Suppliers: Frank W. Joel, Museum Laboratory & Archaeological Supplies Ltd. (Butvars).
            Cairn Chemicals Ltd. (Polyvinyl butyral M.150).
POLYVINYL CHLORIDE, HOT-MELT COMPOUNDS (*see* Vinamould and Vinagel)
POLYVINYL CHLORIDE, POWDERS
  Manufacturer: Imperial Chemical Industries (Plastics Div.) Ltd.
  Difficult to obtain in small quantities. Try Frank W. Joel.
POLYVINYL CHLORIDE, TUBING
  A. Gallenkamp & Co. Ltd.
  Can sometimes be bought at ironmongers and household suppliers.
POLYURETHANE FOAM LIQUIDS, BROWN PARTS 1 and 2
  Strand Glass Co. Ltd.
'QUENTGLAZE'
  Quentplass Ltd.
  Frank W. Joel, Museum Laboratory & Archaeological Supplies Ltd.
RESINS—REMOVING CREAM 'ROSALEX 42'
  Jeb Trading Co. Ltd.
RESPIRATORS, INDUSTRIAL
  Collins & Chambers Ltd.
  A. Gallenkamp & Co. Ltd.
RUBBER LATEX
  Clear: Revultex Grade L.R. (prevulcanised) from:
  Bellman, Ivey & Carter.
  Red: A 800 R from:
  Immediate Transportation Co. Ltd.
SANTOCEL, AEROGEL SILICA
  Manufacturer: Monsanto Chemicals Ltd.
  Supplier: Frank W. Joel, Museum Laboratory & Archaeological Supplies Ltd.
SIEVES AND SIEVING MACHINERY
  Endecotts (Filters) Ltd.
  A. Gallenkamp & Co. Ltd.
SILICONE, COLD CURE SILASTOMERS
  Dow Corning Ltd.
  Imperial Chemical Industries Ltd.
  A. Tiranti Ltd. (small orders).

SILICONE FLUID. M.S. 200/350 cs
  Hopkin & Williams Ltd.
SILICONE GREASE 'RESEASIL NO. 7'
  Hopkin & Williams Ltd.
SOLVENTS, IN BULK 'BISOL' RANGE
  British Industrial Solvents.
SPATULAS, STEEL, PLASTER'S SMALL TOOLS
  A. Tiranti Ltd.
  Shops supplying tools for plasterers.
SPEED ENGRAVER MODEL 72 and 172
  Manufacturer: Burgess Power Tools Ltd.
  Suppliers: with pin chucks: Watkins & Doncaster.
                              Frank W. Joel, Museum Laboratory & Archaeological Supplies Ltd.
  Available without pin chucks through many dealers. May be advertised in newspapers.
STAINS, MICROSCOPE
  E. Gurr Ltd.
ULTRASONIC EQUIPMENT (*see also* 'Cavitron')
  Headland Engineering Developments Co.
VINAGEL 116–18
  Egerton Engineering.
  Frank W. Joel, Museum Laboratory & Archaeological Supplies Ltd.
  A. Tiranti Ltd.
'VINALAK' 5909 and 5911
  Manufacturer: Vinyl Products Ltd.
  Small quantities try: Frank W. Joel, Museum Laboratory & Archaeological Supplies Ltd.
'VINAMOLD'. HOT-MELT COMPOUNDS
  Norman & Raymond Co.
  A. Tiranti Ltd.
  Frank W. Joel, Museum Laboratory & Archaeological Supplies Ltd.
WIRE MESH FOR SIEVES
  A. Gallenkamp & Co. Ltd.

The following list of addresses is in alphabetical order by surnames.

Armstrong Cork Co. Ltd, 351 Caledonian Rd, London N.1.
Ash, Claudius & Sons Co. Ltd, 26 Broadwick St, London W.1.
B.D.H., Poole, Dorset BH12 4NN.
Bellman, Ivey & Carter Ltd, 358b Grand Drive, West Wimbledon, London S.W.20.

Materials, Manufacturers, Suppliers

Blagden, Victor, Co. Ltd, A.M.P. House, Dingwell Rd, Croydon, Surrey.
Bottley, Gregory & Co., 30 Old Church St, London S.W.3.
British Industrial Solvents, Tennants Consolidated Ltd, 69 Grosvenor St, London W1X DBP.
Burgess Power Tools Ltd., Sapcote, Leicestershire LE9 6JW.
Cairn Chemicals, Fir Tree House, 32–40 Headstone Drive, Middlesex HA3 5SQ.
C.I.B.A-Geigy (UK) Ltd, Duxford, Cambridgeshire CB2 4QA.
Citenco Ltd, Elstree Way, Boreham Wood, Hertfordshire.
Collins & Chamber Ltd, 197–199 Mare St, London E8 30F.
Cottrell & Co., 15 Charlotte St, London W.1.
Courtin, Marcel, 'Sunnyhill', Givons Grove, Ashstead, Surrey.
D.C.L. (CO_2 Div.), Distillery Lane, London W.6. Sales office: Broadway House, Broadway, London S.W.19.
Desoutter Bros. Ltd, The Hyde, London N.W.9.
D & J Plastic (Croydon) Ltd, 96 Whitehorse Rd, West Croydon, Surrey.
Dow Corning Ltd, Reading Bridge House, Reading, Berkshire.
Egerton Engineering, Murray Rd, Leesons Hill, Orpington, Kent BR5 3QU.
Endecotts (Filters) Ltd, 9 Lombard Rd, London S.W.19.
Fibreglass Ltd, Reinforcements Division, Bidston, Birkenhead, Cheshire L41 7ED.
Fothergill & Harvey Ltd, Summit, Littleborough, Lancashire.
Gallenkamp, A. & Co. Ltd, P. O. Box 290, Technico House, Christopher St, London E.C.2.
G.E.C.-Elliott Mechanical Automation Ltd, Birch Walk, Erith, Kent.
Graham (Builders Merchant's Ltd), Hawley Mill, Hawley, Dartford, Kent.
Gurr, E., Ltd, P.O. Box 53, Lane End, High Wycombe, Bucks.
Headland Engineering Developments Co., Melon Rd, London S.E.15.
Hopkin & Williams Ltd, Freshwater Rd, Chadwell Heath, Essex.
Immediate Transportation Co. Ltd, 23b St Thomas St, London S.E.1.
Imperial Chemical Industries Ltd, (Head office) Imperial Chemical House, Millbank, London S.W.1.
 (Plastics Div.) Bessemer Rd, Welwyn Garden City, Hertfordshire.
 Local sales offices *see* telephone directories.
 Overseas offices apply to head office.
Jeb Trading Co. Ltd, Colville Rd, Acton, London W.3.
Joel, Frank W. Museum Laboratory & Archaeological Supplies Ltd, 9 Church Manor, Bishops Stortford, Hertfordshire.
Leete, A. & Co. Ltd, 129 London Rd, London S.E.1.
May & Baker, Ltd, Dagenham, Essex.

Appendix 2

(Plastics Div.) 23–25 Eastcastle St, London W.1.
Mono Containers Ltd, Malt House, Field End Rd, Eastcote, Ruislip, Middlesex.
Monsanto Chemicals Ltd, Monsanto House, Victoria St, London S.W.1.
Notcutt, Walter P. Ltd, 44 Church Rd, Teddington, Middlesex.
North Hill Plastics Ltd, 49 Grayling Rd, London N.16.
Norman & Raymond Ltd, 112 Stonhouse St, London S.W.4.
Quentplass Ltd, Thorpe Arch Trading Estate, Boston Spa, Humberside.
Scott Bader Ltd, Wollaston, Wellingborough, Northamptonshire.
Shell Chemicals U.K. Ltd, (Head Office) Downstream Buildings, Shell Centre, London S.E.1.
   For local sales offices *see* telephone directory.
Smith, J. & Son (Clerkenwell) Ltd (Metal Wreho), 50 St John's Square, London E.C.1.
Staident Products Ltd, 33 Clarence St, Staines, Middlesex.
Strand Glass Co. Ltd, Head Office & Mail Order, Brentway Trading Estate, Brentford.
   Branches in: Bristol, Birmingham, Southampton, Ilford, Manchester and Glasgow.
Tiranti, A. Ltd, 72 Charlotte St, London W.1.
Transatlantic Plastics Ltd, (3323) Garden Estate, Ventnor, Isle of Wight.
Trylon Ltd, Thrift St, Wollaston, Wellingborough, Northamptonshire.
Vinyl Products Ltd, Mill Lane, Carshalton, Surrey.
Vinatex Ltd, New Lane, Havant, Hants PO9 2NQ.
Visijar Laboratories Ltd, Pegasus Rd, Croydon Airport, Croydon, Surrey.
Watkins & Doncaster, 110 Park View Rd, Welling, Kent.
Whitley, George M., Ltd, Victoria Rd, South Ruislip, Middlesex.

Index

'Abracaps', 151, 222
Abrasive powders, 76, 249
Acetabular facet, 180
Acetabulum, 186
Acetic acid, dilute for development of:
  complete or partial skeletons, 90, 104–10
  cave breccias, fissure fillings and bone beds, 90, 110–11, 131
  fish by transfer process, 84–90
  invertebrates, 115
  vertebrates in the round, 11, 90
    isolated bones, 90, 91–3
    isolated skulls with or without jaws attached, 90, 93–9
    isolated small mandibular rami, 90, 99–102
    isolated but articulated joints, 90, 102–4
Acetone,
  adjustment of S.G. of heavy liquids with, 131
  as a solvent, 9, 10, 11, 14, 28, 52, 83, 98, 131, 146, 155, 157, 222, 223
Acetylenetetrabromide, use in flotation, 131
Acid,
  large scale preparations, 234
  preparations of invertebrates, 114–16, 118, 119
  preparation of vertebrates, 3, 7, 84–114
Acrylic paints, 217, 222
Adhesives, 1, 2, 5, 6, 9, 27, 37, 235
  application to joints, 26
  desirable properties, 16
  gap filling, 14, 24, 25, 28, 81, 178
  identification of, 6–9
  proprietory, 9, 10, 16
  storage of, 231
Aerogel silica,
  used to prevent high gloss on fossils, 11, 143
  used to thicken rubber latex, 54–5, 198
Air,
  bubbles, avoidance in moulds and casts, 196, 199, 203, 207, 225
  diffused, in flotation, 130
  turbine dental drills, 72, 103
'Airbrasive', S.S. White, 75–7, 112, 118
  abrasive powders used with, 76
  glove box for use with, 77, 263–6
A.J.K. dough, 14, 28, 275, 281
Alcohols,
  as solvents, 9, 10, 13, 21, 144, 146, 148, 155, 157, 240
  for 'de-watering' specimens, 21
Alpha-cyano-acrylate, *see* 'Permabond'
Alum, use in casting, 195
Aluminium,
  bar used as template, 171, 185
  foil, use in the field, 50, 51
  foil, use in the laboratory, 22, 66
  oxide powders as abrasives, 76, 77, 249
  sheet, use in casting, 206
  sheet, use in Transfer process, 86
  stearate, 203
Alveoli, of teeth, 97
Ambient temperature, 88

Index

Ambient temperature (*cont.*)
  effect on setting of polyester resins, 87
  effect on setting of silicone cold cure silastomers, 200
Ammonia, 97, 117
  gas in arrest of pyrites breakdown, 140, 141, 143, 148
  solution in casting, 197, 198
Ammonites,
  casting, 202, 226
  repair of giant, 15
  zone indicators, importance as, 105
Ammonium hydroxide S.G. 0.88, 97, 112
  for treatment of pyrites breakdown, 140–2
Animal glue, *see* glue
Ankle joint, *see* tarsus
Antlers, 10
  preservation of, 13, 20
Anvil, 160, 175, 177, 233, 255, 274
Apatite, in fossil bones, 84
Aqua fortis, *see* nitric acid
Aquatic reptiles, 66
Araldite,
  adhesive, 258, 259
  MY 790 in restorations, 15
Articular bone, 95, 96
Asbestos,
  blocks, 254
  boards, 207
  flexible board, 256, 259
  glove, 254
  health hazard, vi, 205, 250
  paper, 262
  powder as base of heat resistant 'putty', 204, 205
  wool, 202
Asphalt, extraction of specimen from, 136
Atlas vertebra, 110, 175, 183, 189
Aveley, elephant, 44, 45, 168
Avian bones, 7

Backing of moulds, 179, 200, 206
Bags,
  collecting, 38, 57
  polythene, 39, 199, 265
  sand, 63
Balance,
  beam, 222
  point of, in mounted specimens, 163
Bandages,
  in casting, 54, 197, 224
  plaster, application of, 2, 42, 47–51, 53, 58
  plaster, removal of, 56, 58, 59–61, 81, 83
Barium chloride,
  in the arrest of pyrites breakdown, 152
  test for sulphates, 150
Barium hydroxide, in the arrest of pyrites breakdown, 152
Barrier cream, 223
Bases, for mounts, 159, 165, 169, 171, 174, 179, 189
Basket, wire, support for wet specimens, 18, 19
Battens, wood,
  use in the field, 43, 46
  use in the laboratory, 190, 127
Battery operated tool for polystyrene, 120, 154, 272
Beaches, fossils from, 21, 147
Beam for supporting specimen during mounting, 170, 179
Bedacryl 122X,
  as a consolidant, 11
  in the arrest of pyrites breakdown, 11, 142, 151
Bedding plane, 43
Beetles, 147
Benches,
  position in laboratory, 231
  types most useful, 231, 233
Bending metal, 175, 177, 183
Benzene, warning of health hazard, 117, 238
Binocular microscope, *see* microscope
Bipeds, mounting, 189
Biuret test for glue and gelatine, 7
Bivalves, casting, 208

Bivalves (cont.)
  development of hinge teeth, 70
Bleaches, to break down matrices, 117
Blocks of matrix, collecting, 40
Blue vitriol, see copper sulphate
'Blushing', in dried consolidants, 238
Boards,
  in the field, 43
  in the laboratory, 61, 65, 145, 149, 207
  for mounts, see bases
Boiling,
  breaking up matrix by, 118
  low pressure, 20
  point of solvents, 237
Bone beds,
  collecting, 40
  extraction of fossils from, 90, 110–11
Bones, collecting:
  comminuted, 37, 91
  labelling, 42
  loose pieces in the field, 41
  plaster bandaging and packing, 2, 42, 47–51, 53, 58
  wet specimens, 39
Bones, conservation and repair of:
  avian, 7, 33
  with much cancellous tissue, 20, 23
  from caves, 110
  reptilian, heavy, 27
  from sands, 21
  wet, 21
Bones, development by:
  chemical methods, 91–112
  mechanical methods, 59–83
Bones, storage after chemical development, 89, 226–70, 272, 273
'Bostik', backing of ostracoderms, 85
Brachiopods,
  casting, 226
  mounts for, 153
  phosphatised, 115
Brain cavity, 189
Brass,
  casting dams, 206
  for small mounts, 157
  softening, for cold working, 158
  wire mesh, 124, 125
Brazing, 171
Breccias, cave,
  collecting, 40
  extracting specimens from, 110–11
  replacement of phosphates by carbonates in, 110
Bristles, as mounts for small specimens, 121
Bromoform,
  recovery, 131
  use in flotation, 130
Brushes, 38, 46, 91, 94, 95, 112, 143
  consolidants applied with, 17, 24, 95
  resin, cleaning of, 222, 223
  tooth, for cleaning specimens, 80
  typewriter, for cleaning specimens, 80
Bubbles,
  in moulds, 196
  in plaster, 196, 199, 203, 207, 225
Budgerigar, feathers, use in developing fragile specimens, 100
Bulb,
  ceramic holder, 257
  home-made holder, 257
Bulk matrix, concentration of samples from, 3, 123–37
Bulk purchase of materials, 237
'Bumping', avoidance in boiling liquids, 137
Burgess Speed Engraver,
  pin chuck for, 73
  description of, 73–5
Burnt Umber, 54, 198, 205, 218
Burrows, molluscan, casting, 55
'Butvar', 190
  B76, 10, 14, 28, 37, 39, 52, 81, 83
  B98, 10, 14, 22, 39, 53, 64, 81, 83, 197, 217
  solvents for, 10
n-Butyl acetate, 21
  as a solvent, 9

n-Butyl alcohol,
  as a 'de-watering' agent, 21
  as a solvent, 22

Calcium carbonate, fossils composed of, 118
Calcium citrate, 113
Calcium hydroxide, to neutralise hydrofluoric acid, 234
Calcium orthophosphate, use with thioglycollic acid, 112
Calcium stearate, additive to P.V.C. moulding pastes, 203
Canada balsam, 243
Cancellous tissue, treatment of, 20, 23
Canine teeth, protecting during development, 93
Cannal coal, extraction of fossil from, 117
Carbon, specimens preserved as, 84, 115
Carbon dioxide,
  emission in acid technique, 84, 87, 113
  propellent gas for the 'Airbrasive', 76
Carbon tetrachloride,
  as a flotation medium, 131
  as a grease solvent, 228
Carbonaceous materials, extraction, 84, 115, 117
Carbonates,
  development of, 118
  effect of removal from bone cells, 85, 120
'Carbowax', see polyethylene glycol
Card, request for work, 236
Cardboard, 179
  use in casting, 206, 211, 216
  use in field, 38, 50
  moulds, 270, 271
  templates, 155, 159, 160, 189
Carnauba wax, 8
Carnosaur, knee joint, 65
Carpal bones, 104
Carpus, 103
  mounting, 166, 169

Cartilage, 166
Casing, water-proof, in chemical development, 95, 103
Casts,
  in exhibits, 193
  making in the field, 53–5
  flexible, 218, 225
  glass fibre resin, 199, 203, 205, 219
  hollow, 225
  in mounted skeletons, 187, 188, 189, 193
  plaster, 198, 199, 203
  reasons for making, 2, 193
  resin, 195, 198, 199, 201, 203, 205, 218
  of specimens before chemical development, 90, 102, 110, 114
Casting,
  methods, 193–228
  from moulds, 215–28
  from natural moulds, 53, 205
Catalysts,
  for cold cure silastomers, 200
  hazards in the use of, 219–20
  for polyester resins, 87, 219
  storage, 220
Caudal vertebrae, 186–7
Caustic,
  soda, 251
  potash, 251
Cave breccias,
  collection, 40
  developing, 110
  earths, 116
  replacement of phosphates by carbonates, 110
Caves, warning against use of solvents in, 40
Caving helmet, 38
Cavitation, cleaning by, 77
Cavities, 77, 84, 144, 154
'Cavitron', 78–9, 89
Cavity,
  cells, for storage of small fossils, 12, 121
  mounts in plaster, 154
  mounts in wood, 154

Cells, in bone, 84
Celluloid, *see* cellulose nitrate
Cellulose acetate,
  mounts, 121, 156–7
  properties, 244
  sheet, 156
  solutions, 9, 146
  test for, 9
Cellulose nitrate,
  solutions, 9
  test for, 9
Cellulose, surgical, 8, 24, 151
  pads for drying specimens, 150
  pads for holding solvents, 8
Cements, *see* A.J.K. dough, epoxy plastic wood and putty, Fibrenyl, 'Permabond', polymethylmethacrylate monomer polymer mixes
Cementum, of elephant's teeth, 24
Centra, vertebral, 32, 174, 176, 177, 188
Cervical vertebrae,
  excavation, 47
  mounting, 182, 188
Cetiosaurus, casting, 208
Cetrimide, use in arrest of pyrites decay, 146
Chalk, 171, 188
  fossils, 13, 76
  french, 53, 54, 82, 211
  precipitated, 97
Charring, of moulds, prevention, 203–4
Chemical, methods of development, 1, 2, 3, 84–121
Chemical welding of plastics, 156, 157, 268
China clay, *see* kaolin
Chisels,
  cold, use in field, 37, 43
  cold, use in laboratory, 2, 31, 61, 63
  forging and tempering, 254–5
Chitin, development, 84, 115
Chlorides,
  role in pyrites breakdown, 139
  tests for, 21
Chloroform,
  for breaking down oil shales, 117
  as a solvent, 11, 137, 156
4 Chloro-*m*-cresol,
  use in arrest of pyrites breakdown, 140, 147
  warning of hazards, 141, 244
'Chlorox', 117
Chopped strand mat, glass, 82
  use in backing pyritic coals, 149
  use in chemical development, 98
  use in glass fibre casts, 220, 271
  suppliers, 277, 281, 282
Chuck,
  dental, 71, 72
  pin, 69, 73
Chute washing, 128
Citric acid,
  to develop invertebrates, 118
  to develop vertebrates, 113
  to keep iron salts in solution, 150
Clamps, in casting and mounting, 170, 176, 183, 185, 188, 223
Classification of vertebrates for acid development, 90
Claw mounts, for small fossils, 159
Clay,
  breaking down, 116
  collecting from, 41, 46
  modelling, 82, 166, 204, 225
  removal from specimens, 81
Cliffs, hazards when collecting from, 38
Climate, affect on collecting, 39, 42
Clinometer, 36
Clips, for mounting fossils, 159, 160, 161, 163, 169, 171, 172, 175, 176
Closed moulds, 225
Clove oil, as a mould inhibitor, 195
Coals,
  breaking down, 117
  cannal, 117
  fissile, treatment of for pyrites decay, 148–51
  Kilkenny, 148
'Cobwebbing' of sprayed consolidants, 18

Index

Cold,
-setting moulding compounds, see polysulphides and silicone silastomers
-setting resins, see Araldite, epoxy and polyester
Collagen, 244
Collecting, 36
 bags, 38, 58
 from hard rocks, 2, 38, 41
 from soft rocks, 2, 46
 tools, 36, 37
 vertebrate remains, 41–54
Colouring,
 casts, 217, 222
 metal work, 172, 186
Colours,
 for addition to polyester resins, 221
 for addition to polyvinyl chloride pastes, 205
 for addition to rubber latex, 198
 for painting casts, 217, 222
Commercial excavations, 41
Comminuted specimens, treatment of, 5, 26, 37, 53
Concentration, of small fossils, 123–37
Condyles, of skull, 175, 183, 189
Cones, fossil, 147
Conodonts, 115
Consolidants, 2, 9, 24, 231, 235
 optimum strength, 5
Consolidation, 1
 definition of, 5
 in the field, 37, 38–41, 46
 methods, 5, 17–25
Contact edges, 26
Copper sulphate,
 use in Biuret test, 7
 properties, 250
Corals, development by ultrasonics, 78
Corks, in mounting, 162, 188
Cotton,
 floc, 97
 rags, 93
Cotton wool,
 disadvantages as packing, 39, 120
 in mounting specimens, 153
Cracks, filling, 23, 24, 92, 93, 148, 205
Cresol, 4-chloro-*m*, see 4-chloro-*m*-cresol
Crocodiles, long snouted, development of skull, 28, 66
Croid, 204
Crystals, on acid preparations, 88 91
Curing, temperature of polyvinyl paste, 204
Cusps, of teeth, 31

Dams, use in casting, 198, 201, 204, 206, 211, 215
Data, preservation of, 2, 43, 58, 59, 104
Decay,
 of old consolidants and adhesives, 6
 of pyrites, 139–52
Decomposition products of pyrites, 139
Deer, antler, see antlers
Dehydration of specimens, 21
De-ionised water, 7, 13, 20, 24, 37, 92, 105, 118, 127, 131, 150, 232
Dental,
 brushes, 71, 80
 burrs, 71, 72
 casting compounds, 196, 199
 cutting wheels, 71, 72
 drills, 71, 103
 drills, airturbine, 72, 103
 hammer, 61, 62, 72
 handpieces: percussion, 71
  Power's mallet, 72
  rotary, heavy duty, light duty, 71, 80, 151
  Stensiö hammer, 72
 machine, 71–3
 mallet, 72
 mirrors, 65, 95, 144
 picks, 65
 plaster, 215
 probes, 144, 254
 rubber impression compounds, 199
 scrapers, 36, 46, 94

Dental (*cont.*)
  suppliers, 276, 280, 281
  tools, 71, 93
  wax, flexible, 108, 207
Dentine, in elephant tusks and molars, 22, 24
Deserts, fossils from, 80
Desoutter Pneumatic Power Pen, V.P.2, 79–80
Detail, in casts, 193, 195, 218
Detergents, 6, 78, 98, 223, 268
  as separators in casting, 201
  solution to extract plant remains from soils, 130
Development,
  chemical, 2, 84–121
  defined, 59
  mechanical, 2, 59–83
  modern tools for, 71–81
  of thin or narrow structures, 64
  traditional methods, 61–71
  of small fossils, 68–71
Diacetone alcohol, 10, 21, 157, 240
Diamond,
  dental burrs, 71, 103
  dental discs, 271
  dental saws, 31, 71, 98, 99, 109
  writing pencil, 135
Diastema, 161
Dibutylphthalate as a plasticiser, 203, 205
Diffusion grids, sections as supports in chemical development, 91
Dimethyl ketone, *see* acetone
Dinosaur,
  bones, 116
  skull, 110
Dinotherium skull, 273
Disarticulation, by acid, 102
Discs, intervertebral, 176
Disintegration of pyrites, 103
Distal end of bone, 172
Distilled water, 20, 37, 118, 127, 131
Distortion, correction in *Scelidosaurus* bones, 109
Dolomite powder, used in Airbrasive, 76

Dorsal vertebrae,
  excavation, 47
  mounting, 179–80
  spine, 176
Double saucepan, for heating,
  gelatine, 194
  hot melt compound, 202
  polyethylene glycol 4000, 67
Dow-Corning, cold cure silastomers, 200
Dowelling,
  bones, 28
  tusks, 29, 30–1
  wooden, 80, 260
Drawing, recording the position of specimens, 41, 102
Drills,
  dental, 71, 103
  electric, 29, 233
  masonry, 29
'Duplit', casting compound, 196
Dust,
  extraction hood, 77, 83, 151, 232, 271
  hazard, 2

Echinoderms, preparation of, 13, 76, 118
Electrically-heated,
  bucket for hot melt compound, 202
  melting pot for P.E.G. 4000, 256–9
  tool for P.E.G. 4000, 259–63
Electro-magnet, use in separating small fossils, 132
Elephant,
  foot, of Aveley, 44, 45, 168
  teeth, 10, 22–4
  tusks, 10, 22–4, 29–31
Elgin Sandstone fossils, 114
Embedding,
  cones in plastic, 148
  specimens for casting, 209–11
Emulsions, *see* polyvinyl acetate
Enamel of teeth, 31
Enlarged casts,
  from natural rubber, 198
  from silicone rubber, 201, 227

Epiphyses, replacement after acid development, 104
Epoxy resin, 14, 29, 111
  adhesives, 16
  casts, 55, 195, 218, 219
  plaster casts, strengthening with, 217
  'plastic' wood, 15, 128, 145
  'putty', 15, 128, 143, 145
  varnishes, 190, 130, 234
Equipment, field, 36–8
  laboratory, 4, 230–5
Ethyl acetate, 11
  as a solvent, 32, 101, 109, 112, 120, 144, 155, 205
Evaporation, 88, 94
Excavation,
  commercial, 41
  from hard rock, 2, 38, 41
  from soft rock, 2, 46
Exothermic, reaction of,
  epoxy plastics, 151
  polyesters, 87
Extraction hoods, 77, 232, 234
Eyepiece, see microscopes
Eyes, protection of, 38, 111, 115

Fabric, pre-shrunk cotton, use with rubber latex, 197
Feather,
  shafts used as needles, 95, 100
  vanes used as brushes, 100
Femur, 103, 105, 170, 180, 186
  human, 20
Ferrobacilli, inhibition of, 140
Fibre glass,
  backing for specimens, 82, 111, 149, 150
  casts, 195, 198, 201–3, 205, 218, 219, 226, 232
  mounts, 169, 189
  tray making, 270–1
'Fibrenyl' dough, 14, 28, 155
Fibula, 103, 170
Field,
  data, 41, 58, 104
  labels, 47, 57–8
  numbers, 55, 58, 60
  packing, 38, 47
  removal of packing, 2, 59–61
  tools for use in, 36–8, 46
Filler pastes, for rubber, 198
Filter,
  cloths, 125
  colour, 68, 70
  paper, 130
Fin rays, 85
Fins, 85
Fire, retarding agents, for casting resins, 151, 219, 222
Fish,
  scales, 68
  skeletons, 76
  skull, 112
  Transfer method applied to, 85–90
Fissile coals,
  pyrites decay in, 148–51
Fissure fillings, 40, 110, 111
'Flash' line on casts, 208, 223, 224, 225, 226, 227
Flexible,
  asbestos board, 256, 259
  casts, 218, 225
  light guides, 95
  metal dam, 206, 211
Floor polish, as a separator, 271
Flotation, methods, 118, 130–2
Foam,
  polyurethane, use in field, 2, 50, 51
  use in laboratory, 66, 192
  rubber, 62, 65, 66, 70, 120, 264, 265
Foil,
  aluminium, 50, 51
  'Melinex', 82
Foramen magnum, 18, 66, 93, 97, 162, 189
Foramina, 32
Foraminifera, methods of extracting, 115
  separation by flotation, 130, 131
Foreign matter,
  in casts, 216, 223
  in joints, 26
Formic acid, 11

Formic acid (*cont.*)
  use in preparation, 111, 112
Fossil,
  beetles, 147
  cones, 147
  fish, 76, 80, 84, 85–9, 112
  grubs, 119
  insects, 55, 119
  mammals, 31–3, 56, 104, 226
  plants, 55, 117, 130, 147
  reptiles, 14, 27, 28, 46, 50, 52, 60, 65, 66, 68, 72, 78, 79, 81–3, 93, 95, 99–120, 201, 203, 208
  seeds, 146, 147
  wood, 13, 119
Fossils,
  excavation of damp, 39
  porous, 5, 21
Fracture edges, 26
Fractures, major, 61
Frames,
  casting for mounts, 189
  plastic for small fossils, 121, 266–70
Friable specimens, 2, 5, 18, 26, 37, 80
Fumes, precautions,
  acid, 3, 234
  corrosive, 3, 232, 234
  inflammable, 1, 232
  toxic, 1, 234
Fungicide, 21

Ganoid fish, development, 80
Gap-filling adhesive, 14, 24, 25, 29, 81, 145, 178
Gas chamber for pyrites breakdown, 141, 142
Gastropods, casting, 208, 226
Gault, fossils from, 139
Gel coat, polyester, 97, 219
  application to moulds, 220, 227
  mixture, 219, 220
Gelatine,
  consolidation with, 6
  moulder's, 194
  pouring temperature, 195
  properties, 244
  test for, 6, 7
  use in moulding, 194, 218, 226, 228
Girdles limb,
  collecting, 41, 47, 53
  mounting, 179, 180, 189
  preparation of giant marine reptiles, 60, 81–3
Glass window, use in Transfer method, 87
Glass-fibre,
  casts, 195, 198, 201, 202, 203, 205, 218, 219, 226
  loose chopped, 220
  'mothers', 224
  strand mat chopped, 82, 98, 149, 222, 271
  surfacing mat, 220, 226
  woven mat, 53, 81
Glenoid fossa, 186
Gloss, reduction of on specimens, 10, 11, 143
Glove box, instructions for making, 263–6
Gloves,
  asbestos, 254
  leather, 38, 155
  rubber, 83, 111, 115, 223
Glue,
  animal, consolidation with, 6
  brush, 98, 275, 281
  moulds, 194
  removal from specimens, 7
  repair of damage by, 7, 33
  test for, 6, 7
Glycerol (Glycerine),
  additive to gelatine, 194
  removal from specimens, 147
  storage of specimens in, 146
'Glyptal', 88
Gravels, 41, 124
Grids, to support specimens, 91, 93, 99
Grinding,
  techniques, 72
  tools, 256
  wheels, 71, 72, 256
Grubs, extraction from fossil wood, 118

Index

Gum tragacanth, 12, 244
Gun cotton, *see* cellulose nitrate
Gutta percha, 193

Haematite,
  chemical development, 112, 113
  mechanical development, 72
Hammers, 2, 56, 61
  blacksmith's, 233, 254
  bricklayer's, 62
  club, 37
  geological, 37
  piton, 37
  rubber, 28
  sculptor's, 62
Handpiece,
  'Airbrasive', 75, 263
  'Cavitron', 78
  dental, *see* dental
Hardboard, 199
  uses in development, 67, 68, 94, 101, 259
  template in mounting, 174
Hardening, plaster casts, 217
  specimens, 5, 17–26
'Hardy' pick, 36
Heat resistant 'putty', formula, 204–5, 206
Heat sinks, for protection of fossils in photography, 270
Heating iron, low voltage, 22, 101, 259–63, 272
Heavy liquids,
  adjustment of S.G., 130, 131
  flotation methods, 111–18, 130–2
  recovery, 130, 131
Hessian, use in plaster bandaging, 56
Heterodontosaurus, development of skull, 72
Hinge of bivalves, 208
Hinge teeth, of lamellibranchs, 70
Hollow casts, 225
Hominid skulls, mounting of, 162, 165
Hoods, extraction, *see* extraction hoods

Hot,
  melt compounds, 201–3, 208, 212, 219, 232
  plates, 67, 202
  water heaters, 232
  wire, battery-operated, 120, 154, 272
Human skull, mounting of, 162
Humerus, 103, 170, 186
Humidity, relative, *see* relative humidity
Hydrochloric acid, 234
  to destroy bone, 113–14
  to develop chitin, 115
  to develop silicified fossils, 55, 114, 119, 129
Hydrofluoric acid,
  dangers in using, 126, 248
  neutralising, 234
  uses, 4, 113, 115
Hydrogen peroxide,
  storage, 117
  uses, 116
Hypo, *see* thiosulphate
Hygroscopic properties,
  of glycerol, 146
  of morpholine, 144

Ichthyosaur skulls, disarticulation of the bones, 99, 120
Ignition tube, use in casting, 211
Ilia, 105, 179
Ilium, 179, 180
Immersion, consolidation by, 18
Implosion, in cavitation cleaning, 77
Impregnation, double, 20, 25
Impression compounds,
  dental rubber, 199
  mixing board for, 199
  polyvinyl chloride, 53, 162
Industrial methylated spirit, 13
  use with bromoform, 130, 131
  as a solvent, 10, 13, 144, 146, 148, 155
Industrial respirators, 2, 18, 83, 143, 151, 271
Inert liquids, for storage of pyritic fossils, 146

Ink,
  removal from paper, 245
  waterproof, 57, 102, 104
Insect remains,
  in fossil wood, 119
  in nodules, 55
Intervertebral discs, 176, 188
Invertebrates, 36
  casting, 208
  developing, chemically, 84, 114–21
    mechanically, 2, 59–83
Iron blocks, as supports in development, 62
Iron pyrites, 103; *see also* pyrites
Ironstone nodules,
  acid development, 115, 118
  mechanical development, 79
Ischium, 179
Isomantle, 135
Isopropyl alcohol = Isopropanol, as a solvent, 10, 14, 39, 64, 81
Ivory, 22
  casting, 202
  conserving, 22–4
  fragmentation of fossil, 23

Jaw, lower,
  development of, 65
  mounting of, 163, 189
  removal from skull, 90, 96
Jaws, mount for small, 266–70
Jet, conservation of, 22
Joint,
  development methods, 102–4
  disarticulation of, 90, 102–4
  dry, 26
  making a, 26
Joints, articulated, 90, 174
Junction line, in moulds and casts, 208; *see also* flash line
Jurassic dinosaur, development with Power Pen, 79
Jute floc, 197, 206, 245
  base for gap-filling cements, 14, 29, 81, 143, 145, 166
  rubber latex compound with, 97, 108

Kaolin, 14, 250
Karroo, condition of fossils from, 95
Kerosene, *see* paraffin oil
'Keys' in moulds, 212–15
Kilkenny coals, treatment of, 148
Knives,
  use in field, 46
  use in laboratory, 52, 60

Labels,
  conserving, 13, 57–8
  detachment from specimen, 59, 205
  field, importance of, 40, 57–9
  linen, 58
  original, 25, 59
Lamellibranchs,
  casting, 208
  development of hinge teeth, 70
Latex, *see* rubber
Lead strips, 127
Lias, fossils from, 55, 139
Lighting of laboratory, 230
Limb girdles, 60
  casting, 208
  excavation, 41, 47, 53
  mounting, 179–80, 186, 189
Limbs, mounting, 173, 186
Limestones,
  collecting from, 41
  dissolving, 115, 129
Linen,
  labels, 57
  rags, skull supports in acid preparation, 93
Lingual, surface of teeth, developing, 95
Liquid moulding materials, pouring, 197, 200, 206, 207, 216, 225
Liquids heavy, *see* heavy liquids
Lithological evidence, preserving, 119
Loam, consolidating, 16, 40
London clay, fossils from, 139, 147
Long bones,
  collecting, 47
  developing, 104
  mounting, 174

Index

Lower jaws, *see* jaws
'Lucite', *see* Polymethylmethacrylate, 12
Lumbar vertebrae, 176, 178, 188

Magnets, separation of specimens by, 132
Male mould, technique of casting, 169
Mammal,
  skull, 56
  teeth, 31, 32, 99–102, 226
  young bones, 104
Mandibular rami,
  development of small, 99–102
  removal from skulls, 96
  mounting, 184
Manus, 47, 165, 174, 186
'Maraglas' A 655, epoxy resin, 15
Marble, simulated, 15
Marcasite, decay of, 140
Marine reptiles, 81–3
Matrix, removal of, 1, 2
Mechanical,
  methods of development, 2, 59–83
  modern tools for, 71–80
'Melinex',
  clear film, 82
  mirrorised, 65, 95
Melting point,
  of casting compounds, 202
  of polyethylene glycol wax, 67, 246
  of polythene, 246
Mesh,
  nylon, 99, 198, 205
  plastic, 111, 125, 142
  sizes of sieves, 124
  wire, 124
Metal,
  mounts, 155, 157–60, 170–89
  supports, 172, 175, 176, 178, 180
  types used in mounts, 157
Metapodials, mounting, 166
Methylated spirits, 242
Methylene chloride, removal of wax with, 8
Methyl ethyl ketone, 9, 96, 101, 157
  diluent for polybutylmethacrylate solutions, 11, 88, 91
Microbe theory, in pyrites breakdown, 139
Microfossils,
  extracting and concentration, 123
  ultrasonic development of, 78
Microscope,
  binocular, 32, 68, 88, 91, 94, 100, 102, 126, 231
  eyepiece, 68
  long arm, 68, 231
  objectives, 68
Minerals, separation from fossils, 123–38
Miocene beach deposits, USA, 55
Modelling clay, use in casting, 204, 205, 206
Modelling tools, 207, 211
Mollusca, 129
  casting, 226
  extraction by Soxhlet apparatus, 136–7
  mounts for, 153
Molluscan burrows, casting in the field, 55
Monel metal, 125
Morpholine, use in pyrites decay, 144–6, 148
Mould box, 200, 201, 203, 205, 207, 211
Moulding materials,
  pouring, 196–206
  small tool for, 273–4
Mould inhibitors, 195
Moulds, 193
  closed, 225
  making, 206–17
  multipiece, 203, 208, 223–7
  natural, 53, 54, 85, 114, 199, 204, 205, 218
  single piece, 206, 216, 219
Mounting fossils for exhibition, 1, 2, 57, 153–92
  metal mounts for small fossils, 155, 157–9
  metal mounts for complete skeletons, 173–89

Mounting (*cont.*)
  metal mounts for isolated limbs, 165–74
  metal mounts for skulls, 160–5, 175, 183
  plastic mounts, 155–7
  Sizer's method, 154, 155, 187
  on horizontal surfaces, 153, 154, 163
  on vertical surfaces, 154–60, 163
Mounts, storage, 266–70
'Mowital', 10
Mud, removal from fossils, 81
Multipiece moulds, 203, 208, 223–7

Nares, 18, 66
  filling before casting, 205
  protection in acid preparation, 93
Natural moulds,
  casting from, 54, 85, 114, 199, 204, 205, 218
  collecting, 53
Needles, 8
  development with, 69, 71, 88, 91, 93–6, 100–3, 142, 144
Neural arches, 32, 107, 175, 188
  preparing, 64, 65
  used to contain spinal support in mounts, 187–9
Neural spines, 64, 65, 105, 174, 179, 188
Neutralisation,
  of hydrofluoric acid, 234
  of polyurethane foam containers, 52
  tanks, 234
Newspaper, use in field, 38, 42, 57
Nitrocellulose varnish, 139, 244
'Nitromors', solvent for small expoxy joints, 15
Nodules,
  acid preparations, 104–10, 115
  treatment in the field, 55
Non-solvents, to prevent resolidification of dissolved solids, 80
Numbering,
  during preparation, 102
  in the field, 55, 58
Nylon,
  floc, 14

  mesh, uses of, 99, 125, 198, 205
  properties of soluble, 245
  soluble as a consolidant, 12

Ochre, yellow, 93, 166, 221
Oil shales, extraction of fossils from, 117
Old Red Sandstone, 84
Olive oil, separator in casting, 82, 154, 166, 190, 195, 196
Orbits,
  protection in casting, 205
  protection in development, 66, 97
Organic peroxides,
  dangers, 219, 220
  storage, 220
  use with polyester resins, 87, 219
Ornament, of fossils, 39, 68, 70, 76, 84, 207
Ornithischian dinosaurs, 72
Oscillators, ultrasonic, 77
Ossicles, skin, 109
Ossified, tendons, 109
Ostracoderms, development, 84, 114
Ostracods,
  concentration by flotation, 130
  reaction to ultrasonic treatment, 78
Ovens, 67, 91, 94, 142, 149, 150, 162, 192, 202, 225, 231
Oxalic acid, 118, 245

Packing,
  field, 2, 38, 39, 58, 59
  materials, 38, 39, 47
Pads,
  for clips in mounts, 172
  porous, uses in removal of salt from fossils, 21, 150
Palaeobotanical techniques, 56, 114, 237
Palate,
  development, 94
  protection in mounts, 161, 162
Palatine bone, 120
'Panning', method of concentration, 128

Paper,
  asbestos, 262
  conservation of, 13
  covering for benches, 25, 145
  labels, 36
  packing, 39, 58
Paper,
  pulp, 24
  rope, 82
  silicone treated, 33
  thimbles for Soxhlet extractor, 134
  tissue, 38, 97, 107, 120, 149, 150, 199
  use in plaster bandaging, 42, 53
Paraffin,
  liquid, 146, 147
  oil, 198, 201, 228
  wax, 8, 86, 245
Paste, polyvinyl chloride moulding, 114, 202–4, 205, 218, 224, 225, 228
Paxoline, 257
Pectoral girdle, 32
  collecting, 52
  mounting, 180, 189
P.E.G., see polyethylene glycol
Pelvic bones, casting, 208
Pelvic girdle, 32
  collecting, 52
  mammalian, 179
  mounting, 179, 186
  reptilian, 52
  Scelidosaurus, 109
'Permabond', 15
Permeable matrices, methods to breakdown, 116
Peroxides, see hydrogen peroxide and organic peroxides
Perspex, 12
  mounts, 121, 155
  solution as an adhesive, 11, 32, 109, 144
Pes,
  importance of field photographs, 47
  mounting, 165, 174, 186
Pestle and mortar, use in extraction of microfossils, 123

Phalanges,
  mounting, 166
  unguinal, 166
Phenol, 195
Phosphates, 115
  replacement by carbonates in bones from caves, 110
Photographs,
  as aids in preparation, 90, 105
  in mounting skeletons, 174
Photography,
  in the field, 41, 47
  in the laboratory, 105, 198, 269
Picks, use in the field, 46, 56
Piece moulds, see multipiece moulds
Pins, use in mounts, 153
Pipettes, polythene,
  use in acid preparation, 88, 92, 100
  use with consolidants, 17, 24, 144
  method for making, 253
Plant remains,
  treatment in the field, 55
  from sea shores, 147
  isolation from soils, 117, 130
  pyrites decay in, 146, 147
Plaster of Paris, 32
  backing for moulds, 198, 206
  bandages, application, 2, 42, 47–51, 53, 58
  bandages, materials, 42
  bandages, removal, 56, 59–61, 83
  blocks for exhibition, 154, 189–92
  blocks for flat specimens, 66, 82
  blocks, removal of, 66
  blocks as mounts for manus and pes, 166–9
  casts, 195, 196, 198, 199, 201, 203, 215, 224, 228
  dental, 215
  mixing, 43, 215
  'mother', 198, 201, 203, 207, 211, 223
  moulder's, 215
  piece moulds, 193
  restoration with, 14, 24, 32, 93, 145
Plasterer's scrim, 42, 56, 206
Plastic, 1
  foam, 51, 59, 149

Plastic (*cont.*)
  mesh, 111, 125
  mounts, 155–7
  sprays, 60, 127
  tubing, 162, 189
  wood, 15
Plasticine, 66, 69
  use in casting, 198, 201, 203–7, 211, 225, 228
  use in mounting, 82, 166, 169, 174, 176, 187, 190
Plasticisers, 203
Plate glass, 190
Plates, of fossil fish, 76, 84
Pleistocene fossils, precautions in moulding, 202
Plesiosaur, 14, 52, 78, 201
  Fletton, 46, 50
  treatment of pyrites decay in, 141
Pliosaurs,
  collecting limb girdles, 52
  pyrites decay in, 141
Podials, mounting, 166, 169
Polishing plastics, 156
Pollen analysis, borings and columns for, 55
Polybutylmethacrylate,
  use as a consolidant, 11
  use in acid preparations, 88, 91, 93–7, 99, 100, 105, 107, 111
  use in arrest of pyrites decay, 11, 142, 145
  removal from specimens after acid development, 92, 102, 105
  see also 'Bedacryl' *and* 'Vinalak'
Polyester resin,
  backing for pyritic coals, 149, 150
  casts in field, 55
  durability of, 90
  gel coat, 97, 98, 220, 221, 227
  glass fibre casts, 169, 195, 218, 219, 226
  mounts, 82, 169, 189
  properties, 252
  solid casts, 195, 218, 226
  use to harden plaster casts, 217
  use in 'transfer' method, 85–90, 115

  storage, 233
Polyethylene glycol 4000,
  application to specimens, 67
  electrically heated tool for, 259–63
  melting apparatus, instructions for making, 256–9
  properties, 13, 246
  recovery from solution, 94
  removal from specimens, 67
  use in development, 64, 66, 68, 72, 86, 94, 101
  use in preservation of water-logged specimens, 13, 22
  use in storage, 269
  use as a temporary mount, 13, 159, 169
  water soluble 'putty', 97, 148, 205
'Polyfilla', use in restoration, 14
Polymethylmethacrylate,
  adhesive, 11, 32, 109, 144
  frames, 100
  monomer and polymer mixes, 12, 32, 92
  powder, 12
  sheet for mounts, 155, 266
  solvents for, 11
Polystyrene,
  coating in acid preparations, 112
  expanded, shaping, 272
  use in exhibition, 154, 159
  used in storage, 120, 272
  fire-proofed, 272
Polysulphide moulding compounds, 198, 199
Polyurethane,
  foam in the field, 22, 50, 52
  in the laboratory, 66
  in mounting, 162, 192
  toxicity, 52
Polyvinyl acetate, 10, 23
  emulsion as adhesive, 7, 8, 10, 28, 64, 166, 273
  emulsion as consolidant, 10, 18–21, 37, 39, 40, 91, 102, 105, 109, 148
  removal of gloss from dry film, 10, 102
  solution as adhesive, 10

Polyvinyl (*cont.*)
  solution in arrest of pyrites decay, 140, 142, 145, 151
  solution as consolidant, 10
Polyvinyl alcohol, 12
  as an adhesive, 12
  as a consolidant, 12
  as a separator, 66, 108, 198, 202, 204
Polyvinyl butyral,
  as an adhesive, 10, 37
  as a consolidant, 10, 37, 39
  see also 'Butvar' *and* 'Mowital'
Polyvinyl chloride,
  colouring, 205
  flexible casts, 218
  hot melt compounds, 202, 208, 212, 219, 232
  impression compounds, 53
  moulding pastes, 114, 203–5, 224, 225, 228
Polyzoa, developing, 118
  resemblance to joint surfaces of reptile bone, 91
Porosity,
  of paper, 39
  of specimens, 5, 8
Porous pads, *see* pads, porous
Pouring, moulding and casting media, 197, 200, 206, 207, 216, 225
Powder colours, 145, 217, 221
Powder, washing, for flotation, 130
Power tools, *see* Desoutter Pneumatic Power Pen and Burgess Speed Engraver
Pre-accelerated resins, 219
Precipitated chalk, ingredient for water soluble putty, 97
Pre-fil F, fire retardant, 151, 222, 271
Preparation, 1
  chemical, 84–121
  mechanical, 59–83
  of specimens for moulding, 205, 206
Preparator, duties of, 1–4, 235
Probe, ultrasonic, 78, 89
Proboscidian,
  teeth, 22–5
  tusks, 22–5

Protection of teeth,
  in chemical development, 93–5, 98, 100, 101
  in mechanical development, 65
  in mounting skulls, 163
Pterygoids, 110, 120
Pubic symphysis, 179
  position in mounted skeletons, 180
Pubis, 179
Pulverisation of rock samples, 123
Putty,
  heat resistant, 204, 205, 206
  water soluble, 97, 148, 205
P.V.C., *see* polyvinyl chloride
Pyridine, preparation of coals with, 117
Pyrites, breakdown, 3, 11, 139–52
  future approach to, 151
  products of, 139
  specimens subject to, 139
  symptoms of, 141
  test for, 141
  treatments, 139–51
  washing affected specimens, 139, 140, 150

Quadrates, 95, 96
Quadruped, mounting of, 174–89
'Quentglaze', use in collecting wet fossils, 16, 40, 55

Radius, 103, 170
Rag, *see* cotton and linen
Rami, mandibular,
  mounting, 163, 184
  preparing, 65, 99–102
  removal from skulls, 96
Relative humidity, 3, 21, 144
  factor in pyrites decay, 151
Removal of specimens from tablets, 153
Repair,
  of moulds, 201
  of specimens, 5, 6, 26–34
Residue, insoluble, separating from fossils after chemical development, 88, 89, 123–37

Resin,
  cored solder, 158
  removing creams, 223
Resins synthetic, *see* epoxy and polyester
Respirators, industrial, 2, 18, 83, 143, 151, 271
Restoration, of missing pieces, 32, 92
Rheatic bone beds, concentration of fossils from, 131
Ribs,
  articular facets, 185
  collecting, 47, 51
  mounting, 184–6
  repairing, 32
Rocks, hard,
  collecting from, 38, 42–6
  development of specimens from, 64
Rubber latex, 197, 198
  casts, 198
  colouring, 198
  enlarged moulds, 198, 228
  formula for thickening, 55
  moulds, 205, 206, 215
  shrinkage, 55, 197
  use in acid preparation, 93, 96, 99, 107, 111
  use in field, 54
  use to make thin sheeting, 82

Sacking, use in plaster bandages, 56
Sacral ribs, 109, 180
Sacro-iliac joint, mounting, 179
Sacrum, mounting, 175, 176, 179, 183, 187, 188
Salicylic acid, use to prepare small fossils, 118
Salt,
  contributary cause of pyrites breakdown, 139
  removal from specimens, 3, 24, 147
Salts, in acid preparation, 88
Sand,
  trays, 27, 29, 97, 107, 162, 163
  wet, consolidating, 15, 40, 45

Sandstones,
  collecting from, 41, 124
  extraction of fossils from, 84
'Santocel', *see* aerogel silica
Sauropod bones,
  casting, 203
  repairing, 27, 28
'Savlon', use in pyrites breakdown, 146
Scaffolding, laboratory, 27, 170, 179
Scales, fish, 68
Scapula, 109, 180, 186
?*Scelidosaurus*, development from nodule, 105–10
Scelidosaurus harrisoni, type specimen, 65
Scorpion remains, chemical development of, 115
Scutes, of *Scelidosaurus*, 105, 109
Screening, in natural running water, 124
Scrubbing specimens, 81
Sealing oil, for hot melt compounds, 202
Section, vertical, 36
Sedimentation tanks, 126, 232, 234
Seeds, fossil, 146, 147
Separation by heavy liquids, 130–2
Separators, used in casting, 195–7, 200, 205, 207, 211, 212, 216, 219, 225
Serial sectioning, compared to chemical development, 90
Setting time,
  of moulding materials, 200
  of synthetic resins, 87, 217
Shale,
  conservation of, 22
  splitting, 37
Shelf-life,
  of casting compounds, 199, 203
  of resins, 86, 233
Shellac,
  as an adhesive, 8, 74
  in casting, 190, 194, 198, 212, 270
  as a consolidant, 8
  test for, 8

Shells,
  casting, 208
  developing, 70
  extraction from matrix, 114, 116, 129, 130
Shelters, in the field, 40, 42
  hazards, 40
Shock waves,
  danger to specimen, 62
  in hard rock, 64
Shrinkage,
  of adhesives, 16
  of casting and moulding compounds, 54, 196, 197, 200, 219
Shulz solution, 117
Sieves,
  materials, 124, 125
  use in field, 46, 124
  use in laboratory, 111, 118, 124–8
Sieving,
  machinery, 126
  methods, 124–8
Silica,
  aerogel (Santocel 54), *see* Aerogel
  fossils occurring as, 55, 84, 114, 115
Silicified,
  fossils, 114, 119, 129
  matrices, 114, 115
Silicone cold cure silastomers,
  casting and moulding with, 149, 199, 205, 206, 215, 218, 219, 224, 226
  expansion in certain liquids, 201, 228
Silicone,
  fluid and grease, 197, 198, 200, 204, 205
  separators, 82, 97, 107, 197, 202, 205, 216, 218, 225
  storing small fossils in, 147
Silver nitrate, test for chlorides, 21
Sinks, types required in laboratory, 231, 232
Sintered glass, Soxhlet thimbles, 134
Skeletons,
  collecting, 41–53
  developing, 104–10
  mounting, 173–89
  repairing, 31
Sketch,
  field, 41, 47
  for recording position of specimen in washing tube, 135
Skin,
  impressions, 68
  ossicles, 109
Skulls,
  casting, 208
  collecting, 14, 47
  developing, 60, 65, 112
  mounting, 160–5, 175, 183
  repair, 32
  weak points in, 93
Slab mounts, 189–92
Slaked lime, *see* calcium hydroxide
Slate, 37, 63
  slate powder, 218
Slide, cavity, 12, 121
Sloped surfaces, mounting specimens on, 154–60
Small jaws,
  acid development of, 99–102
  mount for, 266–70
'Smooth-on'; *see* polysulphides
Soap,
  laundry, in casting, 194
  soft, as a separator, 82, 166, 167, 201, 212
  solution in flotation method, 130
Soda,
  caustic, 118
  natural, 3
  washing, 118
Sodium,
  alginate, base of casting compound, 196
  bicarbonate, use in Airbrasive, 76, 112
  carbonate, use to break down clays, 118
  chloride, removal from specimen, 3, 24, 147
  hexametaphosphate, uses in development, 116

Sodium (*cont.*)
  hydroxide, use to break down clays, 118
  hypochlorite, use to break down coals and other rocks, 117
  thiosulphate, use of recrystallisation to break down pervious rocks, 116
Soft matrices, fossils from, 80–1
Soft soap, *see* soap
Soils, consolidation of wet, 15, 40, 55
Soldering,
  irons, low voltage, 273
  methods, 158
Solvents,
  chlorinated, dangers of, 239, 240, 242
  inflammable, storage of, 233
  organic, 238–43
  volatile, 232
Solution,
  consolidating, 5–25
  differing rates in acid preparation, 87, 91, 96
  optimum strength of consolidating, 5
Sorbitol, additive to gelatine, 194
Soxhlet extractor,
  description, 133–4
  extracting specimens with, 136, 137
  tubes for use in, 134–6
  washing specimens in, 134, 147
Spatulas,
  how to make a small one for casting, 273, 274
  steel, 28, 54, 149, 167, 190, 205, 222, 224
  wood, 200
Specific gravity, adjustment of, *see* acetone, heavy liquids and methylated spirits
Specimens,
  collecting, 36–59
  comminuted, 5, 26, 37, 53
  concentrating, 123–38
  friable, 2–5, 18, 26, 37

  security, 231
  wet, 21, 24, 39, 55
Spectacles, goggles for use with, 38
Spinal column, mounting, 174–9
Spinning moulds, 224
Splints, use in collecting, 43, 51
Sponges, 60
Spontaneous combustion, warning of possibility, 220
Spray,
  gun, 17
  laboratory, 17
  paint, 17
  plastic, 60, 127
Spraying specimens, 17, 21, 23, 146
Stages in mounting a skeleton, 173–89
Stains, microscope, 217
Stapes, 110
Starr Carr, 21
Steel,
  mild, supports in mounts, 162, 171
  points, 67, 68, 69, 73
  silver, 254
  tempering, 255–6
Steniosaurus, skull of, 66
Sternebrae, 187
Sternum, mounting, 187
Stiffening ridges in glass fibre, *see* paper rope
Storage,
  boxes, 266
  in fluids, 147
  of inflammable substances, 233
  of materials, 232
  of specimens, 2, 119–21
  tubes, for small teeth, 266
Stratified samples, 36, 125
Stratigraphic position of specimens, 36, 41, 125
Strengthening plaster casts, 217
Striking platform, for splitting stone, 109
'Stunning blow', 109
Sulphur, in pyrites breakdown, 139
Sulphuric acid, product of pyrites breakdown, 139

302 Index

Support, for specimen during development, 62, 63, 64, 65, 66, 67, 69
Surfacing mat, for glass fibre casts, 220, 226
Surgical,
 cellulose, 8, 24, 150
 gauze, 197, 200, 206, 215
Sutures,
 as guides to repair, 32
 preservation of, 120
Syn-dibromoethane, flotation technique, 131

Tables, heavy duty, 232, 234
Tablets wooden, in exhibition, 153
Tank,
 brick-built for acid development, 234
 improvised for immersion of large specimens, 23
 neutralisation, 234
 sedimentation, 232
 ultrasonic, 77, 231
Taps,
 for screws, 169, 176, 233
 for water, 231
Tare weight, 222
Tarsal bones, 104
Tarsus, mounting, 103, 166, 169
'Tears' of consolidants,
 avoidance, 20
 removal, 21, 24
'Teepol', a separator in casting, 201
Teeth,
 alveoli, 97
 canine, protecting in acid development, 93
 casting, 196, 199, 202, 226
 cusps, 31
 developing, 65, 70
 elephant's, 10, 22–4
 hinges, of lamellibranchs, 70
 mammal, repair of badly shattered, 31, 32
 protection while mounting, 163
 reptile, protecting in acid development, 93, 95
 storage of small, 121, 266
 wear facets, 94, 100
Temperature,
 ambient, 88, 151
 pouring of moulding compounds, 195, 202, 228
Templates use in mounting specimens, 185, 189
Tendons, ossified, 109
Tent, polythene in laboratory, 18, 21
Terylene,
 floc, 14
 foil, 82
 mesh, 11, 125
Tetrabromoethane, in flotation methods, 131
Tharrhias, 89
Thimbles, Soxhlet, 134
Thiobacilli, inhibition of, 140
Thixotropic, 53
 agents, 55
 resins, 97, 221
Thoracic vertebrae, 179, 180
Thymol, mould inhibitor, 21
Thyoglycollic acid, solvent for haematic matrices, 112
Tibia, 103, 170
Tiles, polystyrene,
 use in exhibits, 154
 use in storage, 272
Tin-plate, for small mounts, 157
Tissues, other than bone, 68
Titanium dioxide, 198, 205, 218, 221, 251
Toluene, 70, 201
 solvent, 10, 11, 142, 145, 151
Toluol, *see* toluene
Tools, 231
Toxic hazards, 1, 3, 53, 130, 132, 137, 230
Traditional development, 61–71, 104
Tragacanth, gum, 12, 244
Transducers, ultrasonic, 77
Transfer method, 85–90
Transformer,
 low voltage, 258, 272

Transformer (*cont.*)
  valve heater, 258
  variable, 263, 273
Translucency of rubber latex, 198
Transverse processes, 65, 174
Trays,
  cardboard, 31, 270
  glass fibre, uses, 91, 93, 103
    instructions for making, 270
  porcelain, 112
  sand, 27, 29
Trilobite, flexible cast of, 218
Trylon resin,
  EM.301, 86
  GC 150 PA, 98, 221, 271
  WR 180, 98, 151, 221, 271
Tube,
  glass, 134
  hard glass, 211
  ignition, 211
  polythene, 17, 253
  polyvinyl chloride, 262
Tubes, rigidity of, 164, 180
Tunnelling, warning against, 57
Tusks,
  composition of, 22, 23
  conservation of, 22–4
  repair of, 30–1
Tweezers,
  swan-necked, 101, 269
  vacuum, 31, 33, 34, 101, 269

Ulna, 103, 170
Ultrasonic,
  development, 77, 89
  equipment, 2, 77–9, 231
  warning against using certain solvents in, 78
Undercuts,
  in field specimens, 56
  in moulds and casts, 194, 200, 201, 206
Undercutting in the field, 43

Vacuum,
  cleaner, 265
  desiccator, 19
  impregnation, 19, 92
  lines, 231
  probe, 34
  pump, 34
  taps, 20
  tweezers, 31, 33, 34, 101, 269
Vapour, phase inhibition of bacilli, 140
Varnish, epoxy, 190, 234
'Velcro', mounts for small specimens, 154
Ventilation of,
  caves, 40
  field shelters, 40
  laboratory, 1, 203, 223, 230
Vertebrae,
  atlas, 175, 183, 189
  caudal, collecting, 47
  cervical, collecting, 47
  dorsal, collecting, 47
  lumbar, collecting, 47
  sacral, collecting, 47
  thoracic, 179, 180
Vertebral column, 105
  establishing height from ground, 179
  methods of mounting, 174–9
Vertebrates,
  casting, 208
  collecting, 41–53
  consolidating, 17–25, 61, 63, 64, 69
  developing chemically, 84–116
  developing mechanically, 65
  limbs, mounting, 165–74, 186
Vertical surfaces,
  casting, 53, 54
  mounting on, 154–60
'Vinagel' 116,
  in the field, 53
  in mounting skulls, 162
'Vinalak' 5909 and 5911, *see* polybutylmethacrylate
'Vinamold', 202
'Vinamul' N 9146, polyvinyl acetate emulsion, 10
Viscosity of solutions, 5
Voids, in P.E.G. 4000, 67

Wall mounts, 143–5
Warping,
  avoiding, 20, 23
  correcting, 109
Washing,
  chamber, 234
  powders as flotation media, 130
  specimens,
    during arrest of pyrites decay, 139, 140, 150
    during and after chemical development, 88, 91, 93, 100, 101, 104, 111, 112, 115, 116, 118
  tubes for use in Soxhlet extractor, 134–6
Water,
  bath, 135, 137
  -logged specimens, 10, 21, 39, 201
  -mixed moulding compounds, 196
  -soluble adhesives, 6, 12, 13, 153
  -soluble 'putty', *see* polyethylene glycol 4000
  -soluble wax, *see* polyethylene glycol 4000
  supply, 127, 231
Water-proof,
  casing, 96–9, 103
  ink, 57
Wax,
  carnauba, 8
  dental, 108, 207
  mixtures, 8, 217
  natural, 8
  paraffin, 8
  polish, 271
  removal from specimens, 8
  water soluble, *see* polyethylene glycol 4000
Wealdon clay, chemical development of fossil from, 116
Wear facets, on teeth, *see* teeth
Welding, 171, 180, 183, 187
  chemical of plastics, 156, 157
Wet paper, use in field, 56, 82
Wet specimens,
  treatment in field, 55
  treatment in laboratory, 21, 24
Wetting agents, *see* detergents
Window glass, use in transfer method, 87
Wire,
  baskets, 18
  mesh, 125
  mounts, 155
  sieve, 125
Wood,
  battens, 43, 46, 127, 190
  fibres, 14
  fossil, 119
  frames, 111, 190
  waterlogged, 13, 201
  wet, 22
  wool, 62
Wooden,
  stocks for heavy specimens, 170
  tablets, 153
Work card, 235
Wrist joint, disarticulation of, 103

Xylene, 70, 243

Yellow ochre, 93, 166, 221

'Zeelex', casting compound, 196
Zygapophysis, 65, 105, 176, 188
Zygomatic arch,
  in field packing, 56
  in mounting, 161